Samuel M Miller

Notes of Hospital Practice

Part I.

Samuel M Miller

Notes of Hospital Practice
Part I.

ISBN/EAN: 9783337161255

Printed in Europe, USA, Canada, Australia, Japan

Cover: Foto ©berggeist007 / pixelio.de

More available books at **www.hansebooks.com**

NOTES

—of—

HOSPITAL PRACTICE

PART I.

PHILADELPHIA HOSPITALS.

Selected and Arranged by

SAMUEL M. MILLER M. D.

PHILADELPHIA, PA.
SAMUEL M. MILLER, M. D., Publisher.

PREFACE.

The following diagnostic and therapeutical notes have, many of them, been condensed and collated by me from my reports of clinical lectures and notes of hospital practice, which have appeared originally in the *New York Medical Record, New York Hospital Gazette, Boston Medical and Surgical Journal, Philadelphia Medical Times, Philadelphia Medical and Surgical Reporter, Cincinnati Clinic* and *Scientific American;* while others are now published for the first time. Every effort has been made to secure accuracy, and it is hoped that the "Notes" will prove of service to the busy practitioner, for whom they are mainly intended. THE PUBLISHER.

GENERAL DISEASES.

SOME INTERESTING POINTS IN THE DIAGNOSIS AND PROGNOSIS OF TYPHOID FEVER.

The case was that of a sailor, admitted to the hospital on the 27th of January, who had been in good health until four days before his admission, when he complained of chilliness, of fever, and of nausea, but of no headache. His nose bled profusely, and his bowels became very loose. Upon his admission *his face was singularly flushed,* and he had a severe pain in his back. His temperature was $104\frac{1}{2}°$ F., his pulse 92, and his respirations 24 to the minute. Nothing could be detected in the condition of the lungs to account for the heavy flush on his face. Upon examining the urine it was found to contain granular hyaline casts and bladder epithelium. It was re-examined, with the same result.

The man remained in the same condition, with morning remissions and evening exacerbations, and with a few bronchial râles in his lungs, until the afternoon of the day after his admission, when profuse epistaxis supervened, and the characteristic rose-colored spots appeared on his abdomen, which grew swollen and tympanitic. *Still there was no headache.* On the evening of January 31st the man's temperature was 103° F. Between January 27th and February 1st there *was never a difference of more than one degree between morning and evening temperatures.* On the morning of February 1st the pulse was only 84, and the respirations 20 to the minute. The tongue was of the characteristic appearance—dry, cracked, reddish at spots, devoid of coating, varnished-looking. The typical spots on the chest and abdomen were slightly raised, and disappeared upon pressure. There was some gurgling in the right iliac fossa, and a moderate amount of abdominal distention. The bowels, after admission, were easily controlled by a single opium suppository daily.

On February 1st the face was still flushed. The breathing was rather harsh, and there were a few dry râles in the lungs. Still no headache, and intellect clear.

Dr. DaCosta, in his examination of the case on February 1st, and remarks upon it, developed some points of much novelty and interest.

The first sound of the heart he found to be very feeble, and there was most marked throbbing of the vessels at the root of the neck. He considered the case to be different from the great majority of cases.

He wished to lay great stress upon the presence of albumen in the urine upon the eighth day of the disease. The case would have to be very closely watched. The presence of the albumen might be explained in either one of two ways—(1) there might have been pre-existing disease of the kidneys as a complication of the fever, or (2) the typhoid fever had produced the disease of the kidneys. If the latter alternative were the true one—and it so seemed to him—the case was a very grave one, for the albumen was noticed as early as the fifth day of the disease. *Early albuminuria, as a symptom, never occurs in the course of typhoid fever unless the case is to be a very grave one.* Albumen is quite commonly found in the urine of typhoid fever patients in the third week of the disease. *The slight difference between morning and evening temperatures so early in the attack was another bad sign.*

Furthermore, the first sound of the heart was thus early altered. Alteration in the first sound of the heart does not usually occur until late in the course of the disease. *When the heart is affected early, it becomes a warning.*

In closing, Dr. DaCosta wished to call attention to the existence of flushed face, *without any disease of the lungs. It always was enough to raise suspicions as to the nature of the disease, especially when accompanied by great throbbing of the vessels at the root of the neck.* This fact had struck him many years ago, and, upon entering a sick room and finding these coincident symptoms, he used to make a rough diagnosis of typhoid fever at once, without any further examination.

All these symptoms being as they were, it was determined to shape the treatment accordingly. Up to February 1st, the man had been taking f ℥ iij of whiskey daily. This quantity was at once increased to f ℥ v. Together with this, gtt. x of muriatic acid was given every four hours. The daily distributed dose of quinia was gr. x. The man's diet was very carefully regulated, consisting principally of beef-tea and milk. Diarrhœa was checked by opium suppositories. The patient was sponged morning and evening with tepid water.

Feb. 20*th.*—The man is now convalescent, having been carried through the attack by careful treatment. The albuminuria has disappeared.

SOME INTERESTING CASES OF TYPHOID FEVER, AND THEIR TREATMENT.

The following cases occurred among the sailors of the Russian steamers which were built and repaired during the summer and fall of 1878, on the Delaware river at Philadelphia:

Out of 550 sailors, 30 were attacked with different grades of typhoid fever. During the months of September, October, November and December, these cases were brought to the German Hospital. They were all seen by the attending physicians, Drs. Turnbull, Woodbury and Cohen.

With the assistance of Dr. Hermann and of the Russian physicians, it was determined to ascertain the cause of the outbreak. The majority of sick sailors came from one steamer, and as their drinking water was different from that of the officers, the first clue as to the cause of the disease was thus found. Examining into this, right in the immediate vicinity of this steamer a privy was found, a large part of the excrement from which found its way into the water from which the sailors drank. On the other steamers this state of things was not found to exist, but the excrement and dirt on board was thrown over into the river, from which they at first obtained their drinking water. Surmising this to be the cause of a few cases of the disease they resorted to another source for their water, which caused the disappearance of the disease. The interesting cases brought into the hospital were:

B., æt. 23, robust constitution, entered the hospital after feeling out of sorts for five days. From his own statement, the prominent symptoms were, intense headache and great heat in the evening; did not seem to know the condition of his bowels, nor could any other symptom be obtained from him. On entering the hospital he presented the following condition: Great headache, high fever, temperature 103° A. M., 104° P. M., pulse 90 to 100, skin hot and dry, great thirst, bowels constipated, tongue heavily coated with a dark-brown fur—deeply fissured, tenderness in right iliac fossa, no tympanitis and no eruption. He remained in this condition for seven days; on the eighth day he became delirious, necessitating his being tied with shackles; he refused to take any nourishment and would not respond to any command. His temperature rose to 105° and his bowels were moved with an enema. On the twelfth day his delirium became active, wild, and after a heroic dose of morphia hypodermically, he appeared rational the next morning; his bowels were moved every third day with an injection; never once having the slightest tendency to diarrhœa; the injections acted very rapidly. On the fourteenth day after admission his temperature fell to 104° P. M., 103° A. M.; less headache, the skin still hot and dry; tongue heavily coated with a dark fur—deeply fissured; no rose-colored spots could be found on the body, and his bowels were still constipated. Late in the evening signs of approaching delirium seemed evident and he was given a large hypodermic injection of morphia; he slept well after the injection of morphia and appeared rational the next morning.

He continued in this condition until the twenty-first day, when his temperature rose to 105°, with great tympanitis, and diffused abdominal tenderness. He lost consciousness and died the same evening.

Autopsy.—Body was much emaciated. Lungs consolidated posteriorly, (hypostatic pneumonia.) Spleen enlarged and softened. Peyer's patches were in different stages of inflammation, some swollen, others sloughing and others having nothing but peritoneum for the floor of the ulcer. No perforation was found. This was what would be called one of the grave forms of typhoid fever, though the temperature did not reach above 105°. He was systematically fed, but the emaciation in the last few days was very great.

The following is a good illustration of what has been called an insidious, latent, or ambulatory form:

G., æt. 37, was working at his usual duties as a sailor, noticing nothing abnormal except a slight diarrhœa, which caused so little disturbance as to pass unheeded. As he himself expressed it, his bowels were opened only once more than usual during the day. He had no evening heat, no lassitude, nothing, in fact, that would lead one to think he had typhoid fever. All of a sudden he felt a sharp pain in the abdomen, which, in a few hours grew unbearable; his belly became very tympanitic and tender to the touch; his temperature was 103°, P. M.; rose-colored spots were found on the abdomen. He now presented all the symptoms of acute peritonitis and was treated with opium. He died in twelve hours after his admission.

Autopsy revealed ulceration of Peyer's patches in the third week, with a large perforation in an ulcer near the ileo-cœcal valve; recent lymph over peritoneum and bowels, indicative of peritonitis. Spleen enlarged and softened; other organs found healthy.

In addition to these, there were some very mild cases, which would be classed under the head of abortive variety. Three cases of this kind presented themselves, with the following symptoms and course:

Some lassitude; headache, which increased in intensity in

the evening; temperature 102°, P. M., 100°, A. M.; pulse, 90 to 100. Very little diarrhœa, stools thin in consistency and of a yellow color. Tenderness over right iliac fossa. In ten days three of these cases showed characteristic rose-colored spots on abdomen, one having numerous spots on both extremities; one, also, had considerable enlargement of the spleen.

In the beginning of the third week the temperature reached the normal and all symptoms of fever declined. Bowels became regular, headache disappeared, tongue clean, &c., &c.

These cases were kept under observation for weeks afterwards to see if any untoward symptoms appeared, while at the same time they were kept upon a liquid diet. Four weeks afterwards they all left the hospital fat; one stouter than he was previous to the fever.

The following was a very grave case: R., æt. 22, entered the hospital about the eighth day of the disease, with very high fever, pulse rapid, 120 to 124, temperature 106°, P. M., 104°, A. M., appetite capricious, tongue heavily coated and very dry. Bowels very loose, having as many as twelve stools per day, resembling pea-soup in appearance; his headache was intense; ringing in the ears and flashes of light before the eyes. This condition of things went on for four days, and then the characteristic eruption appeared. In the meantime his diarrhœa had been checked by the treatment. The temperature remained the same, and his headache was increasing. On the thirteenth day, *i. e.*, from the beginning of the disease, he was suddenly seized with an active, wild delirium, which required constant restraint and watching, in order to prevent him from hurting himself.

He passed his urine and fæces in bed, unconscious of what he was doing. His urine was examined and found highly colored, acid on reaction, but no albumen. Two days after the setting in of the delirium he became comatose, and died shortly afterwards. The temperature the last two days was 105°.

Autopsy showed the characteristic lesion of Peyer's patches, with congestion of brain and lungs.

The following seemed to be a recurrent attack of typhoid fever showing an abortive course: C., æt. 25, had all the characteristic symptoms of the fever—diarrhœa, headache, tenderness in right iliac fossa, rose-colored spots and high temperature. In the latter part of the fourth week the temperature began to decline, and in five days reached the normal. It remained normal for ten days, and he presented the condition of one convalescent from typhoid fever. Then the temperature began to go up for five days, until it reached 103°, P. M., and remained stationery for six days, with a slight morning remission. First, the diarrhœa and headache returned, and two days afterwards the eruption re-appeared. After thirteen days of relapse the temperature went down, and in three days reached the normal. He was very soon convalescent.

Out of thirty cases, twenty-one had diarrhœa differing in severity; some having nine to twelve stools, others two to five, per diem; four, constipation, (going five days without a passage from their bowels,) which was quickly relieved by an enema. In the remaining six, diarrhœa and constipation occurred alternately. Sometimes their bowels would be regular for four or five days, and then diarrhœa or constipation would set in, and so on throughout the course of the disease.

The eruption appeared in twenty-one cases—the majority showing it on the abdomen. It appeared, in a few cases, at the same time, on the back. Two out of the thirty showed spots on both extremities. The number of spots varied; in some as many as twenty-five were counted—in others only from two to five could be seen. The two having spots on the lower extremities had the largest number. Tenderness in the right iliac fossa was found in a good number of cases, but it was also absent where undoubted signs of typhoid fever existed. Delirium showed itself in one-third of the whole number of cases. Some had very high temperatures, and others had very mild temperatures. Those cases having very high temperatures, with active delirium, generally proved fatal.

Three cases had a temperature of 106°, but had no delirium during the whole fever.

Lung complications occurred in the majority of the cases, in the form of slight attacks of bronchitis, or of pneumonia, or pleurisy—the former predominating. Hemorrhage from the bowels occurred in one case, which recovered. As regards the prognosis, the most unfavorable signs were high temperature, active and wild delirium, and severe diarrhœa. In two of the fatal cases there was a very prominent symptom—i. e., a quivering of the extremities upon the slightest motion. When asked to put out the tongue, it came slowly, tremulous and uncertain. These two patients were not able to protrude the tongue to its fullest extent, or when they did so it was, quickly retracted, on account of an inability to keep it out. Such manifestations were of unfavorable omen.

THE TREATMENT OF TYPHOID FEVER.

Dr. Pepper regards the specific follicular catarrh of the intestines as of great importance in the determination of the treatment. He holds that there are a number of remedies which exert a powerful influence upon this catarrh. The first of these is the nitrate of silver, which reduces the size of the enlarged follicles, relieves the inflammatory engorgement, and allays the hyperæsthesia of the nerves. So too do carbolic acid and the subnitrate of bismuth. He prefers the nitrate of silver, and only substitutes carbolic acid in its place when some putrid element is present. The nitrate of silver is given in quarter-grain doses, four times daily. Opium in pill form is combined with the silver when the diarrhœa is excessive. From one-quarter of a grain all the way up to one grain of opium is given thrice or four times daily. If the bowels are constipated, belladonna is substituted in the place of opium.

Milk is regarded as the best diet during the stage of catarrhal inflammation. The milk is diluted with lime-water if the curd appears in the stools. Two pints of milk and lime-water, mixed, are given in the course of the twenty-four hours.

The poisoned state of the blood is controlled by means of quinia and salicylic acid. The latter has been found to be very valuable in this condition as a disinfectant. Quinia is given to the amount of not more than twelve grains in twenty-four hours.

Temperature is kept down by preventive measures rather than by the cold bath, which is regarded as a *dernier ressort*. When temperature runs up in spite of drugs, the whole body is sponged every two hours—the sponges being squeezed out of a mixture of water and bay rum, at a temperature of from 60° to 80°. If the patient's temperature still runs up he is wrapped in sheets wrung out of cold water. The only case in which the cold bath is used is in the first ten days, where the temperature rises above 103°, and is not to be controlled by milder measures.

Stimulants are not administered to patients under the age of puberty, as a general thing. They are only thought to be demanded by some one or more of the following indications, viz.: (1.) Ataxic nervous disturbances. (2.) Profound asthenia. (3.) Circulatory disturbances. (4.) Dry and brown tongue with sordes. The milder forms of stimulus, such as wine whey, are always used at first. When demanded, whiskey is given with lime-water—the latter being added to prevent coagulation—in the proportion of f ℥ ss. each of whiskey and lime-water to every f ℥ iij. of milk.

Relapses are treated as first attacks. Hemorrhage is managed by absolute rest in bed for twenty-four hours, and by the administration of opium to produce absolute rest for the alimentary canal. Acetate of lead is sometimes combined with the opium in the shape of suppositories. Ergot is also very useful. The food allowed is small in quantity and liquid.

Dr. Cohen finds that large doses of quinia rather increase the diarrhœa and headache. His patients are sponged with vinegar and water, and abundant ice is given them to suck.

Oil of turpentine in 20 gtt. doses every hour or two in mucilage, he has found to act most beneficially. Morphia is given hypodermically in half-grain doses where the delirium is active.

Dr. William H. Bennett reduces the temperature and strengthens the heart by 10 gtt. doses of digitalis thrice daily.

Dr. Louis Starr controls diarrhœa by starch and laudenum enemata.

PLEURO-PNEUMONIA FOLLOWING TYPHOID FEVER—DEATH FROM LARGE PLEURITIC EFFUSION AT END OF SEVENTH WEEK, WITH PEYER'S PATCHES STILL UNHEALED.

George Ross, æt. 26, born in Massachusetts, a sailor, single; admitted October 30th, 1877; always enjoyed good health, and denies venereal disease; had been sick four weeks at sea before admission, the principal symptoms being prostration and some fever. He was in a condition of slight hebetude when admitted. It was difficult to obtain much history, but the captain of his ship stated that there had been no diarrhœa, epistaxis, or actual delirium. Respiration was slightly jerking, but the lungs expanded well and no abnormal sounds could be detected in the chest. He had no cough; the abdomen was not tympanitic, and the marks of some kind of plaster were noticed in the hypogastric and iliac regions.

Temperature upon admission was 101°, the pulse 101, and the respirations 24 to the minute. The patient was restless and irritable and required watching at night, although there was no marked delirium. He seemed rather dull and did not complain of anything. Bowels were moved two or three times daily after admission. The urine was light colored, cloudy, sp. gr. 1010, no albumen and no sugar.

November 2d.—Patient found to have left-sided pneumonia; dullness at left base posteriorily, where there is also impaired respiration, coarse crepitation, with bronchophony and bronchial breathing above and whiffling in respiration at line of dullness. These signs were not found in the front, where the percussion note was clear on both sides, and the respiration vesicular. There was no cough and no expectoration.

The nails showed checking of growth in a ridge running across the nail at about one-third the distance from the matrix to the free border. He was ordered turpentine stupes, quinia (gr. xij) daily, and a mixture containing gtt. v of the tincture of digitalis and gr. x of the citrate of potassium every four hours. The impulse of the heart was moderately extended, but not forcible. The first sound was murmurish and soft, but there was no murmur.

November 3d.—Physical characters much the same; still no cough. There is impaired breathing, and some fine râles can be heard at the right base posteriorly, with some loss of resonance. The dullness on the left side is clearing up; the respiration anteriorly is rather exaggerated; the amount of whiskey increased to f ℥ vj daily.

November 5th.—Decided dullness at right base posteriorly, with signs of pneumonic consolidation of lower half of lung. The left side is clearing. Ordered carbonate of ammonia (gr. x) every three hours.

November 7th.—Considerable dullness of right side anteriorly; is troubled with hiccough upon the slightest exertion; this continues nearly all night; has a slight cough occasionally; he is restless at night.

Treatment continued; blister to right chest anteriorly; spirits of chloroform (f ʒ j) administered four times a day; takes ten ounces of whiskey daily; tongue coated and dry; no cardiac murmur; a little faint crepitation at base of heart, but the patient's condition prevents him from holding his breath for the purpose of accurate diagnosis.

November 8th.—Tongue tremulous; has a little dry cough and troublesome singultus. Ordered plenty of milk, beef-tea, eggs, &c.

November 9th.—Looks better, but still very weak; still considerable dullness over the upper part of the right lung, where respiration is imperfect and feeble, with slightly prolonged expiration. The whole condition seems to give the idea of non-expansion or of collapse. The pulse has better volume; tongue cleaner, moister and slightly coated, but is less typhoidal in its appearance. There is less hebetude.

November 11th.—Hiccough better; still taking ten ounces of whiskey daily, and carbonate of ammonia every two hours; respiration rather feeble at the right base, with coarse crackling above it; tongue rather dry, with white coating, but less dry than before; no albuminuria. Basham's mixture (f. ℥ ss. s. t. d.) now ordered. There is some effusion at the right base.

November 12th.—Twitching of the facial muscles in sleep; respirations 23 in sleep and 36 awake; has very little hiccough at present.

November 13th.—Better to-day than yesterday; appearance brighter; urine normal; pulse has rather more volume; there is a sound of friction at the right base.

November 14th.—About the same; temperature rather high; respiration quite feeble, with some prolongation of respiratory murmur at the right apex, with still some dullness; fremitus on right side feeble; the evidence of effusion at the right base persists. Whiskey increased to f. ℥ viij, and Basham's mixture given four times daily; quinia, gr. xvj, daily, and carbonate of ammonia, gr. x, every four hours.

November 15th.—Impulse at apex of heart is not in the normal position, but is a little outside of the linea mammalis; the sounds are indistinct, yet well defined; the heart is displaced by pressure from the right side, but its beat is distinctly felt; the lower part of the right lung moves a little in respiration. Ammoniae carb. to be stopped, and Basham's mixture to be given every three hours.

November 18th.—Is passing water freely; apparently improving; respiration, though feeble at the lower part of the right lung, is again distinctly vesicular and is less dull, but there is still impaired resonance at the right apex; the apex beat of the heart retains its distinctness, and is seen as well as felt in the sixth intercostal space, one and one-half inches to the outside of the nipple; a slight quiver can be seen in the space above; there is slight friction at the left base; respiration well sustained on the left side, low down, but it is rather harsh.

Died this afternoon.

A post-mortem examination reveals an immense effusion in the right chest, with condensation of lung by pressure; heart pushed to left; no active sign of pneumonia.

In the ileum were found patches of Peyer in an ulcerated condition, one having sloughed out, leaving the sub-mucous coat with a healthy, granulating surface.—*Dr. DaCosta.*

INDICATIONS AGAINST PARACENTESIS THORACIS.

The case had been in the wards for some time, with the history of an attack of pleurisy following exposure. When the patient was first admitted in November, examination revealed an old right-sided pleurisy, with some evidence of abscess of the right lung. Subsequently the signs of pleuritic effusion developed rapidly, and it was very evident that there was considerable effusion in the lower part of the chest.

The question which arose was whether, with the evidence of a right-sided pleurisy, which remained rather stationary, together with the suspicion of tuberculous disease, resort should be had to aspiration, or whether the endeavor should be made to get·rid of the effusion by medicinal means. When the patient was first admitted he had already had the effusion for several months.

Upon thinking the case over, and considering the strong probability of disease of the lung itself, though masked; finding also no marked irritative fever, and having, therefore, no reason to suppose that the chest was full of pus, Dr. DaCosta concluded to try and get rid of the effused serum by medicinal means, and determined not to tap the chest. The result justi* fied the conclusion reached.

On February 1st the dullness still remained low down in

the right chest. The voice was transmitted from all other parts of the lung. So, too, with regard to the vocal fremitus. The breathing was also fuller and deeper.

The treatment consisted principally in the administration of the tincture of the chloride of iron, with acetate of ammonium. Occasionally a Dover's powder was given at bedtime. The food was generous, and counter-irritation was frequently made with iodine or blisters.

From the good results already shown, Dr. DaCosta was confident that this treatment ought to be persevered in.

The patient was ordered in future a tablespoonful of Basham's mixture four times daily, and a tablespoonful of cod-liver oil thrice daily, on account of the suspected latent disease of the lungs.

The case was regarded as proving that it is never wrong in old cases of pleural effusion to give a fair trial to medicinal means first; and never to try tapping until we are quite sure that all other modes of relief are of no avail.

The *three* points particularly suggested and emphasized by the case were—(1) that we should be guided rather by the effects of an effusion than by the time it has lasted; (2) the value of Basham's mixture and repeated counter-irritation in the treatment of chronic pleurisy; (3) the possibility of tuberculous disease of the lung as a co-existent factor is always an additional reason for not tapping, since surgical interference should never be attempted when this complication exists.

PNEUMONIA.

The routine treatment of pneumonia in the wards of the Pennsylvania Hospital consists in the internal administration of from eight to twelve grains of quinia daily, together with a moderate amount of nitrate of potassium and of the tincture of digitalis every two or three hours. Plenty of stimulus is administered. In a case recently under treatment, Dr. DaCosta gave in place of the nitrate of potassium a teaspoonful of the spirits of ammonia in water every three hours, as an alkali. This use of ammonia (as an alkali) was so successful in Dr. DaCosta's hands that Dr. Hutchinson tried it in one of his cases with equally good results.

THE TREATMENT OF ORGANIC HEART DISEASE.

In those instances where there is a maximum amount of cardiac force, with a minimum amount of valvular lesion, cardiac sedatives are regarded by Dr. Pepper as the remedies *par excellence*. Veratrum viride, aconite, the bromide of potassium, and other bromides are given in small and continued doses. The need of cardiac sedatives has been found to be most marked in diseases of the mitral valve, where there is a marked tendency to hypertrophy of the left ventricle.

In these cases the diet allowed is cooling and restricted; circulatory and nervous stimulants are avoided. If the general system is plethoric a saline depurative is administered. The diet, though restricted, is not reducing, *i. e.*, the blood is not reduced in quality by it, though it is of such a kind as to be easily digested.

Where the valvular lesion has been the result of an endocarditis contracted in early life, it has often been found possible to accomplish the greatest amount of good by continuous doses of the iodide of potassium. This treatment has often cured young children with hypertrophied left ventricles and mitral disease.

In those cases of heart disease showing impairment of power, and occurring late in life, the most important item of treatment is rest and the avoidance of all muscular effort. Such patients are given beds on the ground floor and never allowed to mount stairs. In bad cases rest upon one floor and in one room is insisted upon.

The question of diet in the treatment of heart disease has received unusual attention in the Hospital of the University of Pennsylvania. The diet is studied in connection with the state of the system. When the digestion is good and the blood not in abundance, the patient is allowed bread, meat, fruits, and green vegetables quite freely. Some such patients have been benefited by a lean meat diet. No patient in this condition can digest oil well. In those cases where the digestion is not good, koumyss, buttermilk, or skimmed milk is given. Where the secretions are scanty and dropsy is present, the diet prescribed is exclusively one of milk. Such patients are not allowed to eat much at a time, but take food frequently and in small quantities.

Where spasms of cough and of dyspnœa occur at night, the patient is only given a small amount of stimulus and liquid nourishment for some hours before going to bed.

(To return for a moment to the question of rest. In some mild cases of heart disease, gentle, moderate walking is strongly advised, but in no instance is needless running hither and thither allowed.)

For the relief of the various congestions consequent upon heart disease, counter-irritants are applied over the affected part. Where nervous and head symptoms predominate, dry cups are applied to the nape of the neck. In pulmonary congestion, muriate of ammonia is given internally in addition to the external counter-irritation. The bromides are used in cerebral congestions. When the stomach is congested, blue mass is prescribed.

When the appetite is poor, the stools insufficient, the liver tender upon palpation, and the secretions of the intestines scanty, blue pills, followed by a saline laxative, is a favorite remedy. Renal congestion is put a stop to by digitalis, together with a saline diuretic.

When the system is in an anæmic state, when the blood is watery, and when it is deficient in red globules, iron is given with advantage. Active plethora is always regarded as a counter-indication to the use of iron. The iron, when given, is administered in the form of a laxative ferruginous water, or a diuretic ferruginous mineral water.

In the treatment of the various dropsies complicating heart disease, cups, blisters, iodine painted on the surface, or iodine with croton oil is used.

In some cases the dropsy is entirely cured by rest and a skimmed-milk diet.

In cases of anasarca the most rapid relief is obtained by the use of jaborandi. Where the heart is so weak that jaborandi cannot be used, resort is had to laxatives, or warm vapor baths.

Ascites is met by saline diuretics; hydrothorax by diaphoretics and diuretics. The patient's whole body is periodically examined physically, to see that no effusion is gaining headway.

If the dropsy becomes otherwise unmanageable, resort is had to operative measures, and the skin is tapped—a number of minute punctures being made in the skin with delicate needles.

Where there is a faulty condition of the nervous ganglia of the heart, associated with the organic disease, digitalis is employed with great benefit. Where one preparation of this drug is not borne by the stomach, another is substituted. Where the separate contractions of the heart are evidently inefficient, and the pulse is weak and small, digitalis has proved itself an unrivaled remedy. The usual dose of the tincture is gtt. x; of the infusion, f℥j; and of digitalin, gr. 1/60 every three hours.

Belladonna has been of service where the heart's action is strong but irregular. Where the heart muscles are weak and passive congestions rife, strychnia and quinia are prescribed.

ALBUMINOID DEGENERATION OF THE KIDNEYS OF PROBABLE SYPHILITIC ORIGIN, WITH OZÆNA AND PSEUDOMEMBRANOUS ANGINA.

The patient was a white man, thirty years of age. When admitted to the hospital he presented the following history:

He had been engaged in the manufacture of morocco for the past fourteen years, with the exception of three years spent in a butcher's shop. His habits had been good, always, except when butchering, when he was intemperate. He had always been a hard-working man, and had been constantly exposed to cold and wet. He had variola and measles when a child, but no other exanthematous disease. He was a married man, and had one child, who suffered constantly from œdema of the feet and face. His (the patient's) father died of paralysis of syphilitic origin, and his mother of dropsy. His only living brother was healthy, but his sister had goitre, palpitation and dropsy.

During the twelve months prior to his admission the patient had had frequent pains in the back and frequent dyspnœa. During the last four months of this time there had been a general condition of anasarca.

The appearances presented by the man upon admission were as follows: His face, neck, feet and body were swollen. His general aspect was markedly anæmic. He was decidedly drowsy, but his intellect was good. His tongue was tremulous, coated, and flabby, and there was a membranous patch upon his uvula. The tonsils were enlarged, and the throat sore. The man's appetite was good, but there was considerable flatulence and pyrosis after meals. His bowels were somewhat costive. Respiration was attended with a loud noise in the larynx. Expiration was prolonged. The glands on the right side of the neck were swollen. He had a bad taste in his mouth, and his breath had an unpleasant smell. There was no abdominal effusion, and his lungs seemed to be healthy. The sounds of the heart were distant and feeble, and the pulse was small and rapid. The apex-beat of the heart could be felt underneath the rib in the fourth interspace; a weak impulse could also be felt in the second and third interspaces at the border of the sternum. The upper border of cardiac dullness was a line drawn from the head of the second rib to the left nipple, and its utmost limit outwards was a line drawn directly downwards from the nipple to the fifth rib. A line drawn parallel to the linea mammalis and running from the fifth rib to the xiphoid cartilage represented the lower border of cardiac dullness, while its limit on the right was the median line of the sternum. The urine passed was dark red in color, and of alkaline reaction, and amounted to sixty-four ounces in the course of twenty-four hours. The contained albumen formed fully one-half its bulk. Its specific gravity was 1006. It contained many granular and hyaline casts.

Dr. Pepper regarded the case as one of a very curious and interesting nature. It was concluded that the kidneys were the seat of albuminoid degeneration, with catarrhal nephritis. The underlying constitutional taint seemed to be undoubtedly syphilitic. It was argued that the angina was an outcome of the systemic disease, but that the ozæna was presumably a manifestation of the catarrhal nephritis.

The case was regarded as an excellent example of those cases of hereditary syphilitic infection in which the poison is so diluted and the patient's constitution so robust that the disease does not show itself until well on towards middle life.

The indications for treatment were thought to consist in (1) the removal of the dropsy by large doses of jaborandi, and (2) in a nourishing and easily digestible diet. Buttermilk, oatmeal gruel, and light broths were ordered as foods.

After the dropsy had largely disappeared minute doses of the bichloride of mercury, with the iodide of potassium, were administered; at the same time cod-liver oil was given in combination with the iodide of iron. The patient was much improved.

IDIOPATHIC PERITONITIS.

If the case is brought into the wards at the very inception of the disease, the patient is bled thoroughly from the arm. If the disease is of many hours standing, Dr. Wood has the abdomen covered with as many leeches as it will hold. After venesection, calomel is administered in doses of from one-quarter to one-half of a grain every hour. In connection with the calomel, opium is given in large doses. Opium induces quiet and prevents the exhaustion consequent upon horrible physical pain. Enough opium is given to keep the patient on the verge of narcotism. It had better be given in liquid form.

In the latter stages of peritonitis, blisters are always employed.

The first thing done, however, when the leeches have been removed, is to apply poultices; whether they be hot or cold makes but very little difference. Where there is a very marked tendency to feverishness, cold poultices are used. If the abdomen is too tender to bear the weight of the ice-bag, light flannel cloths wrung out of ice-water may be used. On the other hand, a warm-water dressing may be employed with advantage in very many cases. Warm water acts not only as a local derivative, but some of it probably oozes through the intervening tissues into the abdomen, and so acts directly upon the inflamed peritoneum as a soothing agent.

After the abdomen has been thoroughly poulticed for two or three days, blisters are used, provided the temperature of the body has not remained high. The blister should not be a small one—eight inches by ten makes a very good size.

When there is any septic element in the disease, quinia is used with great benefit. Generally the stomach is not strong enough to bear it.

The patient must have but very little food in the first few days of the attack. The food which is given is that which leaves the least residuum of undigested matters, and therefore causes the least amount of peristaltic action on the part of the intestines. Milk, in repeated small doses, is the food usually given. At the end of a few days, solid articles are allowed. When there are symptoms of exhaustion late in the course of the attack, beef-tea is given as a stimulant. Alcohol is not only powerless, but even dangerous in the early stages of the disease. A few doses of brandy in the first few days of an attack of peritonitis may produce death.

With regard to the opening of the bowels during convalescence, a purgative or an enema is never used. These bring violently into play all the muscles of the abdomen. Very often there will be a spontaneous movement on the fifth or sixth day without any medicine at all. If there is not such an opening, a small dose of castor oil is given at the end of ten days. If there is retention of urine, the water is, of course, drawn off by means of the catheter.

Great care is had during convalescence from peritonitis to prevent a relapse. No violent or gymnastic exercise is allowed for a long time afterward.

TWO INTERESTING CASES OF VOMITING.

CASE I,—æt. twenty-five. Family history not good. Sister and father both died of phthisis. She herself was always healthy. Began to menstruate at age of seventeen and stopped menstruating at twenty. Since then the menses have been very irregular. She married at the age of eighteen and was a widow at twenty-one. You will, therefore, notice that the irregularity of menstruation has existed not only during marriage, but also before and since.

She comes into the hospital for the treatment of what is apparently a very serious difficulty, viz., she has been vomiting constantly for nearly a year. She has been in the hospital for only a week. She has been vomiting incessantly; has never retained more than one meal during the course of the day for the past year. The vomiting always begins the instant after her meal is over. She does not have much vomiting or nausea between her meals.

A year ago she was stout and healthy, but the vomiting has rendered her thin and pale. Though not so this morning, it is fair to state that she has picked up wonderfully within the last few days, and you will, no doubt, want to know what has been done.

In the first place, I had the vomited matter examined, with but negative results. There were no sarcinæ, nothing but mucous mixed with the contents of the stomach. Occasionally the patient has been disturbed by vomiting mucous in her sleep.

Together with the vomiting you notice that she has a slight, irritative cough. This cough has troubled her ever since the vomiting began. Joined with the cough there is no expectoration.

Before going any further, however, I will examine the gastric and intestinal organs. Her tongue is slightly coated and flabby, and there is some tenderness in the epigastric region and along the spine, particularly at about the middle point of the spine. There is, however, no appearance of a tumor. The soreness in the epigastrium is general, and is not localized in particular spots. The bowels are constipated; the respiratory sounds are normal. There is no albumen or abnormal ingredient in the urine. There is no fever, and the temperature is normal. The urine is acid and of the usual color, with a specific gravity of 1025.

What has been the cause of vomiting? What remedy is it which has stopped the vomiting in three days?

When I first saw the woman in the wards and heard of the incessant vomiting, I first thought it was a case of irritable stomach in a young woman, connected with gastric ulcer. The epigastric soreness, the age of the patient, the appearance of the tongue, the disordered menstruation, the sore point in the spine, all tended in that direction.

I soon gave up the idea, however. Gastric ulcer gives rise to local soreness; here the soreness was general. Gastric ulcer is attended by hemorrhage and pain upon taking food, which was not the case here. The two most prominent symptoms of ulceration were absent. I rejected the idea.

Then there came up a point of experience in my mind— one case similar had happened not long since in my own practice; several I had seen in consultation. In one case the vomiting had reduced the patient almost to the verge of the grave; nothing stayed on her stomach. In her case the irritation was reflected upon the stomach from a uterine malady—a slight flexion. It was not very different from the sympathetic vomiting of pregnancy. Moreover, there was in that case a certain amount of gastric disease—a catarrhal affection of the stomach came on as a complication of the nervous affection. In the light of that experience I began to suspect the same condition here. As a result of examination we found retroflexion of the uterus. The whole case was cleared up at once. It was reflex vomiting with a certain amount of gastric catarrh, lasting for a year, although the woman had taken the greatest care with her diet, etc.

When she was first admitted she was put upon lime-water and milk, but there was no effect produced upon the vomiting. How did we check it? It was not by diet alone.

Again experience came to my assistance, and I determined to try the application of ice to the spine as a systematic treatment, every few hours. The ice was applied and left on until it chilled the patient and made her skin cold. The application was often repeated—as often as she could bear it. Its effects were admirable. No other treatment was necessary. The vomiting stopped almost at once.

You will not always be so successful with this remedy alone. Is there nothing which we can combine with the ice—must we depend upon it alone? By no means. Bromide of sodium in doses of ten or fifteen grains thrice daily is very effectual. It lessens the reflex irritation, and is not rejected by the stomach. An occasional purge by an enema, or some bitter-water, is often desirable. Subsequently, if the case lingers, use blisters to the spine. Do all this irrespective of the local uterine treatment, for the reduction of the uterine displacement will not always stop the vomiting at once—the cause is removed, but not the habit.

To-day I shall introduce a pessary. Already the girl's

diet has been increased, and she is beginning to retain solid food.

What else can we use in such cases to soothe the stomach? Pepsin is very valuable as soon as the vomiting has been stopped. The dose of saccharated pepsin is five grains thrice daily. The diet all this while should be gradually increased; only small quantities of food being given, but these frequently.

This condition is very similar to hysterical vomiting. There is no manifestation of hysteria here. The two states are parallel, but not identical. In an hysterical case the results of treatment would not be so good.

CASE II,—æt. fifty-five, single, comes of healthy family. Her own health has not been bad considering her age and occupation—that of cook. She has suffered from dyspepsia for a long time, together with flatulence and constipation. She is also a sufferer from sick headache, and from a certain amount of pain in her stomach.

Some time after these symptoms appeared, her abdomen began to swell and became painful. She vomited; on two occasions vomited blood. Her general health at the same time has failed, and she has lost flesh, and is, as you see, very pale and anæmic. This morning she vomited blood for the third time.

The treatment shall be based (1) upon the hæmatemesis, and (2) upon the general gastric symptoms. Since she has been in the hospital she has vomited every day. Attending this vomiting there has been a burning pain in her stomach. This pain is not always increased by taking food, and often existed apart from the hours of her meals.

The vomit, a specimen of which I here show you, consists of black coagula—coffee-grounds—the usual character of the vomit of cases of hæmatemesis. The attack of vomiting which she had this morning has rendered her pale and weak.

The disease is plainly shown, by the character of the vomit, to be situated in the stomach; is probably gastric ulcer

Her temperature has risen to 100°, her pulse is feeble and compressible, there is general soreness over the epigastric region, the tongue is dry and but slightly coated. The woman is scarcely in a condition for me to attempt an accurate and minute examination and diagnosis. When the gastric hemorrhage is over, it will be very easy to make out the gastric condition; indeed, I have already told you what I believe it to be.

As regards the treatment of the hæmatemesis, the most effectual remedy is the hypodermic injection of ℥x of the fluid extract of ergot the minute that the bleeding begins. This injection should be repeated if the least symptom of return of hemorrhage appears. We shall keep up this woman's strength by means of such food as eggs, milk, beef-tea, etc., by enema. If there is any sign of heart-failure, brandy is, of course, necessary by enema. Lastly, I shall order several small blisters to be placed over the epigastrium. In this way you see that we are treating and nourishing the patient without putting a drop of anything in her stomach. Should this treatment not be successful, I should order ℥x of turpentine and gr. $1/24$ of morphia in emulsion, by the mouth, every third hour.—*Dr. DaCosta.*

GRAVES' DISEASE.

In the treatment of this disease Dr. Pepper gives the greatest care to the removal of the causes, and towards securing rest, good food, change of scene and entire release from care. The various functions are attended to, and any local disorder in females is removed by suitable treatment. Digitalis has been found to be the most valuable remedy for controlling the functional disturbance of the heart. It is given freely, in doses of from ten to fifteen drops, three or four times a day, and continued for long periods. When anæmia exists, large doses of iron are administered. Most excellent results have been obtained from the injection of diluted solutions of ergotina into the enlarged thyroid gland. The needle is introduced to the depth of half an inch or an inch, and from six to ten minims of a solution containing 96 grains of ergotina to the f℥j of distilled water was injected. Bromide of potassium has been found to assist the iron and ergot in regulating the action of the heart.

INFLAMMATIONS OF THE NASAL PASSAGES, EUSTACHIAN TUBES AND MIDDLE EAR.

A solution of zinc is applied through the catheter. To do this, the catheter is first introduced, and then three or four minims of a solution ($3/5$ gr. to the f℥j) of zinc are dropped into it. The zinc is forced through the catheter into the ear by means of Politzer's bag. In some cases nitrate of silver is applied to the diseased surface by means of a post-nasal syringe introduced behind the soft palate. In old cases of catarrh of the middle-ear, where the secretions have ceased, the attempt is often made to stimulate the membrane. This is done by means of ether. From 8 to 10 gtt. of ether are dropped into Politzer's bag. The patient takes some water in his mouth, and holds it there. A nose-piece is put in his nose, and just as he is swallowing the water, the ether is squeezed through the nose-piece into the passages.

With regard to constitutional measures: where the disease has been hereditary, and has run through many generations, the case will only go on from bad to worse, unless something be done to bring up the general tone. If there be any strumous diathesis, the bichloride of mercury is given internally for a long time, and in small doses.

Dr. George Strawbridge regards the following as a good form of administration of the drug:

R. Hydrarg. chlo. corrosivi... gr. $\frac{1}{30}$.
 Elix. cinchonæ... f℥ss.   M
Sig —To be taken two or three times a day, after meals.

Iron and strychnia are given in pill form. In old people, where there is very decided lessening of the secretions, ten-grain doses of the muriate of ammonia are given thrice daily. Before administering this dose it is dissolved in f℥j of the elixir of cinchona, and this again suspended in half a pint of acid water. Muriate of ammonia, like iodide of potassium, is never administered to the stomach unless in a highly diluted state.

THE NIGHT-SWEATS OF PHTHISIS.

"I have treated the night-sweats very successfully in this and other cases with granules of atropia—$1/80$ to $1/60$ of a grain every night before retiring. This atropia treatment was first started at the Pennsylvania Hospital, and it has been very generally adopted elsewhere, as the best means of checking colliquative sweats, both in phthisis and in other affections; after taking from $1/80$ to $1/60$ of a grain of atropia every night for four or five nights, the sweating is usually entirely checked. It has been objected to this atropia treatment that it produces great dryness of the throat. I have endeavored to counteract this effect by the use of strong lemon-juice, lemonade, gum-water, or slippery elm conjointly with the atropia. Quite recently I have obtained the very best results by means of jaborandi combined with the atropia. The doses of jaborandi must be exceedingly small. I have not yet made trial of this combination in a sufficient number of cases to enable me to make an authoritative statement. All I can say is that I have been able to produce entire toleration of the atropia in every case in which the jaborandi was combined with it. The atropia checks the great drain of the sweats upon the system, and so gives the other remedies a chance to act."—*Dr. J. M. Da Costa.*

CONTAGION IN SCARLET FEVER.

On April 30th, 1878, a very severe case of scarlet fever appeared in the boys' ward of the Children's Hospital. The boy attacked occupied a bed near the centre of the room. Coincident with this was a similar case in the girls' ward. Both of these wards are on the second floor. The cases were at once removed to the upper story, where wards are prepared for such contagious diseases as arise among the house patients.

Within a week previous to his removal this boy had been successively changed from two beds to a third, and immediately afterwards these two beds had been occupied by other boys. No other case occurred in either of these wards. In a ward on the third story, front, were a number of children suffering from whooping-cough, a disease which had run riot throughout the house, but was then upon the decline. The ward adjoining this was the one devoted to the scarlet fever patients. Among the whooping-cough cases, on the 2d of May, a girl manifested the characteristic symptoms of scarlet fever, and was transferred to the adjacent ward. Not another one of these cases of pertussis took the scarlet fever, notwithstanding the fact that the girl already attacked, who was well grown, and was suffering also from keratitis in addition to the cough. She acted as nurses' assistant, and in this capacity repeatedly came in contact with the other children in the ward.

Ten days after the last date, on May 12th, two very mild cases, occupying beds very remote from each other, broke out in the back ward on the second floor. These last were so faintly marked that they might have been overlooked but for the suspicions excited by the prevalence of the disease. On the 1st of June one other case occurred in this ward. On May 4th another case occurred, a boy suffering from a slight deformity, and confined to the "play-room," so-called—a room below the first floor, opening directly upon the playground. The children from this part of the house seldom came in contact with those in the upper wards, and then only in exceptional cases. Twenty-four days after the isolation of this last boy, one other case, that of a girl, occurred in the "play-room."

From these facts it would seem that *contagion* is *not* the most important element in the spread of this disorder. We must rather look to individual susceptibility coupled with similarity of circumstances. Nor has mere physical weakness much to do with this susceptibility. Lying very near most of the cases above mentioned were children suffering from a great variety of troubles, both medical and surgical. Only one or two in an entire ward were attacked. During the time which elapsed from the incipiency of the first to the convalescence of the last case, between eighty and ninety general patients were treated in the house. Only nine of these took the scarlet fever. Of all the patients in the house, but half had had the scarlet fever. The first boy attacked had been in the house seven weeks before the manifestation of the angina, which, in his case, was unusually severe. He could scarcely have come in contact with any one carrying the "seeds of the fever" after entering the house. This case was by far the most severe of all, resulting in death. None of the others died. Each case was milder than the preceding one—an evidence that the most susceptible are those first attacked. To but one case was there the sequela of albuminuria. This child was also suffering from a severe bronchitis, and at the outset of the fever was much emaciated. The presence of the albumen was first detected on the nineteenth day. There was little else to suggest danger, except a trace of blood in the urine. Prompt treatment held this symptom in check, however, and by three weeks' time the blood and albumen had both disappeared from the urine. The bronchitis meanwhile was all gone, and the child improved in color and flesh. Keratitis in two cases was much improved by the fever.—*Dr James H. Hutchinson.*

PATULOUS AORTIC VALVE ACCOMPANYING CHRONIC BRIGHT'S DISEASE.

The patient was found lying unconscious in the street. When brought into the hospital his pupils were contracted and his breathing stertorous. The radial arteries were very tortuous. One-sixth of the bulk of the urine was found to consist of albumen. The fæces were passed involuntarily.

Immediately upon admission the man was given gtt. ij of croton oil in a teaspoonful of olive oil, and a dose of the following prescription:

 ℞ Spts. chloroformi (B. P.).................................gtt. xxx.
   Acid. benzoici..gr. vj.
   Potas. bicarb...gr. xxx. M
 Sig.—Drops ten every two hours, in water.

For sustenance the patient took three pints of milk and one pint of beef-tea, together with f ℨ j. of alcohol, the milk and beef-tea being given in small quantities at short intervals.

Atropia was injected into the eyeball, but did not dilate the pupil sufficiently to enable the visiting physician to examine the eye-ground with the ophthalmoscope.

When examined carefully by Dr. John Forsyth Meigs, on the day after admission, the right side of the patient's mouth was found to be paralyzed. So, too, were the right arm and right leg—the leg less so than the arm. The man was still dull and inattentive. His respirations were 44 to the minute, and his pulse 116. The radial arteries wound along like worms, and pulsated visibly from the elbows to the wrists. Another curious fact in connection with the case was the presence of rounded, hard tumors on the back of his elbows and over his phalangeo-metacarpal joints. One of these tumors was opened, and a very thick, creamy matter escaped, which yielded crystals of tyrosin under the microscope. It was impossible at that time to detect any murmur of aortic regurgitation.

In endeavoring to reach a correct diagnosis in the case, the visiting physician was led at first to regard the case as one of opium-poisoning owing to the marked contraction of the pupils; but this view of the case was invalidated by the facts of paralysis and stertorous breathing. It was finally determined that the case was one of chronic Bright's disease, and the opinion unhesitatingly advanced that further examination would reveal the presence of tube-casts, and that when the more violent symptoms had subsided a regurgitant cardiac murmur would be heard. Mention was made of the statement advanced by Walsh, of London, viz., that the tortuousness of the arteries present in the case is only present in patulous aortic orifice and coarctation of the aorta.

Subsequent events proved this diagnosis to be the correct one.

The patient's comatose condition, dependent upon the uræmic poisoning, was found to yield very markedly to the following prescription, recommended by Dr. George Johnson, of London, viz.:

℞ Scammonii resinæ..gr. v.
   Potassii bitart..gr. xx.
   Zingiberis..gr. viij. M
Sig.—To be administered when needed.

RHEUMATOID ARTHRITIS AND CHRONIC ECZEMA.

James Talbot, white, æt. 41, iron worker. Admitted July 30th, 1874. Twenty years before that time he went to the Isthmus of Panama in charge of a storehouse, and remained there four months, during which time he contracted Panama fever, consisting of chills, fever and night sweat, followed by general debility. He returned to Philadelphia very much reduced in health and has never since recovered from the effects of this disease. Fifteen years ago he was attacked for the first time with inflammatory rheumatism, beginning suddenly in the feet and extending thence to the various other joints of the body. He was confined to bed for nine months, the feet, ankles and hands enlarging and becoming painful. He remained in a partially crippled condition for five years being better and worse from time to time. Seven years ago, while at work excavating a well, he was again attacked with rheumatism, accompanied by great pain in the joints and by general stiffness. Upon this occasion he kept his bed for fifteen months and experienced a most severe attack. He never recovered entirely from this illness, and found himself becoming more and more involved from month to month. Three years ago the disease had increased to such an extent that he was forced to stay in his bed permanently. His feet, hands, and all the larger joints were greatly swollen and disfigured. He now began to get steadily worse, both as regards deformity and suffering. The rheumatic symptoms all became more marked and showed no disposition to leave him. He became entirely unable to use any of his joints and even to turn in bed. He has been on his back for the last two years. During this time he has gradually lost flesh, notwithstanding that his appetite has been good. His bowels have been constipated. He has never had any skin disease of any kind until one year ago when his present trouble began to show itself on his right foot, around the back of the foot and on the instep. The left foot soon became similarly affected. Later the hands and arms were attacked, and by degrees, in the course of six months, the trunk became involved. Patient states that the eruption started about the large toe nail, which loosened and came away, and was followed by a crust formation. These crusts came off from time to time leaving a reddish, discharging surface, and finally all the toe nails became detached. At this time there was no itching, but about eight months ago itching began and was most severe about the legs and arms.

Present condition.—The skin of the face, scalp, shoulders, arms, abdomen, penis, legs and feet, is involved in a low grade of eczema. The disease is diffused in patches of various sizes, and is most marked upon the lower extremities. The skin is boggy and red, giving forth a copious secretion of liquid with a certain amount of blood, which dries, forming extensive crusts. On some parts of the feet there is maceration of the epidermis, showing the *rete mucosum* in an exposed state. The nails of the toes are all in a soft, broken-down condition and loosely attached to the matrix. Upon the calves of the legs are small, abraded patches of papular eczema. The knees also are similarly affected. About either side of the throat at the anterior termination of the ribs are two patches of inflamed skin, resulting from the constant pressure of the hands upon these parts. The hands and fingers are covered with large, yellowish crusts, which can readily be detached, exposing beneath a soft pulpy substance, from which serum and blood constantly ooze. The skin upon the other portions of the body is more or less shriveled, dry and scaly. A very low state of vitality of the cutaneous surfaces is everywhere to be seen. Patient lies upon his back and is unable to move from this position. The thighs are flexed upon the body and the legs upon the thighs. The joint trouble and the contraction of the tendons fixes the limbs in this position. The feet are everted to almost a right angle with the legs, and are immovable. The arms are fixed across the abdomen. The left hand abducted and fingers immovably flexed. The right hand is less markedly distorted, and the fingers are extended. The lower jaw is almost fixed and he is unable to masticate anything solid. Appetite good and is in good spirits. Urine has a specific gravity of 1015, and contains red albumen or sugar. He sleeps pretty well, but is kept awake during the early part of the night by pain in the joints and limbs. Respiration 16, pulse 84. Bowels constipated, was ordered pulv. aloes, gr. j; ext. colch. comp., gr. j; ext., hyoscy., gr. ss., at night, and his body to be washed with castile soap and water, and then sponged with liq. picis alk. f ʒ to Oj, and also an ointment of acid. carbol., gtt. x, ung. zinc. benzoat, ʒ ij, to be applied after the tar wash.

Aug. 8th, 1874.—Has been steadily improving as regards

the eruption, healthy skin in extending all directions, particularly upon the feet, where its advances can be daily noted. Had a despondent nervous attack yesterday. Bowels not moved for three days, so repeated laxative pills. One or more of the joints are moved daily by the nurse. Patient is taken out upon his bed into the open air for an hour or more a day, according to his condition, weather permitting.

Aug. 28th.—Prostrated somewhat by a nervous attack. Pain in limbs, more motion than formerly, skin improving, steam bath administered in bed and continued twenty minutes. Pulse increased from 88 to 115.

Sept. 9th.—Steam bath to lower limbs.

Sept. 22d.—Slept comfortably; better than he has been for two years. General condition improving. Omit tar alkaline wash. Decidedly more motion in joints than formerly.

Oct. 1st.—Eruption changing from squamous to papular and vesicular, particularly upon the lower limbs.

Oct. 7th.—Some little improvement, appetite good. Continue the same treatment and diet, milk punch daily, and ale in evening. External application of green soap and alcohol. Complains of soreness and stiffness in jaws on account of taking cold, some increase in number of blebs on legs.

Oct. 14th.—Motion of joints improving. Blebs not quite so numerous.

Oct. 29th.—Decidedly better. Legs and feet much improved, ordered ol. morrhuæ, f ℥ ss., t. d., and stopped potas. iod. mixture.

Nov. 9th.—Ordered a pill of ferri proto. carb.; gr. ij, pulv. glycr., gr. ij; sacc. alb., gr. iv., t. d. Also an ointment of amyl, ℥ iij, zinci ox., ℥ iij; hydrarg. chlor. mit., ʒ iv, to be applied to limbs and body night and morning.

Nov. 18th.—Blebs very numerous and confined to the calves of the legs. This increased quantity of blebs seems to contradict the idea that they are caused by the potas. iod., for that medicine was discontinued long since. Ordered the iron powders to be stopped.

Dec. 1st.—No improvement. Ordered tr. ferri chlor., gtt. v, in liq. pot. arsenitis, ℳj, thrice daily.

Feb. 2d, 1875.—Patient has been no better for last two months. Feet almost in same condition. No material change except a little lessening in number of bullæ.

Feb. 22d.—Pain in joints is very severe now. No change in condition of skin.

Mar. 15th.—Use tar ointment on feet and legs. Stopped the ointment of calomel.

April 25th.—No change. Ordered a prescription of hydrarg. chlor. corros. and alcohol.

Aug. 24th.—Discharged to-day unimproved. For the past two months he has been taking little or no medicine, and there has been no change in his condition. About August 1st, 1878, he died in another hospital in this city.—*Dr. Louis A. Duhring.*

DILATATION OF THE STOMACH.

The endeavor is made by Dr. Cohen to produce contraction of the walls of the stomach by washing out the ingesta with the stomach-pump, and then rinsing out the stomach with a solution of some disinfectant. The double effect is thus had of cleansing and gymnastic exercise. In inserting the stomach-pump the head of the patient is turned back, and the tube is passed down the œsophagus over two fingers placed in the mouth as a guide. This treatment by the stomach-pump is persevered in for a long time. The stomach is well rinsed out at each sitting. To get rid of remnants of food, emetics are employed. The patient's general tone is sustained by tonics.

ACNE ROSACEA.

Dr. F. F. Maury treated this condition by the local application of a mercury plaster. The alimentary tract is kept thoroughly open by a methodical course of salines, such as Epsom and Glauber salts, and Crab Orchard and Hunyadi János water.

In addition to the mercury plaster as a local application, the following lotion was ordered for the face:

    ℞ Sulphuris sublim... ℨij.
        Etheris... ℨiij.
        Vini frumenti.. q. s.
    Ft. lotio.

The patient was advised not to drink anything but a little red or white wine, and to be careful to refrain from fish and too much meat.

TINEA FAVOSA.

The hair is first pulled out by the roots, the parasites killed, and the scabs poulticed. Before beginning to pull out the hair, however, the whole head is closely shaved. The separate hairs are pulled out with delicate tweezers. These hairs are pulled out in their long axis, otherwise they would be liable to break off short. The poultice is applied immediately after the hairs are pulled out, so that the pustules may be well softened. After applying the poultice some parasiticide—such as the iodide of sulphur, or the bichloride of mercury—is rubbed in. The parasiticide most frequently used by Dr. J. Solis Cohen is that formed by adding gr. ij of sulphur iodide to ℨi of lard.

SCIATICA.

In Dr. Bennett's hands hypodermic injections of one eightieth of a grain of atropia and one-eighth of a grain of morphia directly into the substance of the affected muscle have always afforded immediate relief, although this relief, as a usual thing, is only temporary. In cases of sciatica Dr. Starr has injected the one-eightieth of a grain of atropia into the tissues directly over the track of the painful nerve with manifest benefit. Dr. Wharton Sinkler, in several cases of local rheumatism, has tried the local injection of ether—gtt. x in water. Though this injection generally relieves the pain, yet it is of itself always a most painful operation, owing, no doubt, to the local irritation caused by the ether.

"When there is distinct local inflammation I am accustomed to treat the disease with large doses of iodide of potassium and minute doses of the bichloride of mercury. As there is not much pressure exerted upon the nerve-trunks in this case, the iodide of potassium will suffice. If we desire to cause absorption of inflammatory matters inside the sheath, the best way to do so is by means of severe blistering or by the use of the actual cautery. The actual cautery, in particular, has great absorbent action, and powerfully relieves oversensibility of the nerves. In this present instance I have been using the actual cautery with the greatest success at the points where the pain has been most marked. Another excellent treatment is by hypodermic injections of morphia and atropia right down into the adjacent muscular structures. For this purpose we are in the habit of using here from one-sixth to one-fourth of a grain of morphia and from one-ninetieth to one-sixtieth of a grain of atropia. In employing this formula you must be careful as the disease subsides that the opium habit is broken. Indeed, the formation of such a habit should be guarded against by intermitting the treatment from time to time. In still other cases, again, where the localized pain has been intense, I have derived most excellent results from the hypodermic injection of from eight to twelve minims of chloroform, taking great care to keep the needle out of the way of the arteries. Though incomparable as a temporary destroyer of pain, the effects of the chloroform are not very permanent. Before leaving the subject I may refer to the use of electricity in these cases. Galvanism is very quick, in some instances, to relieve pain. The mode of application should be with the positive pole at the seat of the pam and the negative pole along the nerve-trunk. Where the muscles have wasted to any great extent the faradic current is the best. By some of the above-mentioned agents you will generally be able to cut short the course of the disease."—*Dr. Wm. Pepper.*

HÆMOPTYSIS.

Dr. Starr has injected hypodermically from two to three drops of the fluid extract of ergot in this condition with most excellent effects, always rapidly stopping the hemorrhage.

SACCHARINE DIABETES.

The diet upon which a diabetic patient is placed consists, for breakfast, of eggs, and any kind of meat, except oysters, gluten bread, and tea or coffee, with milk and without sugar; for dinner, tomatoes, lettuce, onions, spinage, string beans, meat, light sour wine, and lemons, or, perhaps, oranges, but none of the sweet fruits; for supper, about the same as for breakfast. None of the starchy foods, no alcohol, and no sugar are allowed.

Among drugs, opium is considered the most valuable by Dr. Pepper. Of this an immense amount can be taken daily, it has been found, without producing any of the symptoms of poisoning. One patient with this disease took seven grains of opium per day. In this case the only bad effect was the production of obstinate constipation. In some cases, even this was unnoticed. The opium, by directly diminishing all the secretions, or more probably by its action on the nerve centres, relieves the excessive thirst and voracious appetite, and reduces the amount of urine and of sugar in the urine. In one case the daily amount of urine was reduced by the use of this drug, from twenty-eight to eleven pints. Ergot, which has been found to act in simple diuresis almost like a specific, is used in saccharine diabetes with much profit, in doses of one drachm of the fluid extract four times a day. Where the skin is dry and rough, jaborandi has been found to be of value, by reason of its great powers of diuresis. The opium and ergot are stopped while the jaborandi is being used.

GASTRIC VERTIGO.

The best treatment has been found by Dr. DaCosta to consist in regulating the diet and in the administration of bitter tonics and of alkaline waters, such as Vichy and Karlsbad. These waters are taken, of course, after meals. The general tone of the stomach is, at the same time, invigorated by the use of bitter tonics before meals. Later on in the course of the disease, iron is used in combination with strychnia. In some cases corrosive sublimate has been found to exert a most beneficial influence upon the gastric and cerebral phenomena.

The hypodermic injection of dialyzed iron has been tried, but the results have not been invariably good, owing to the unreliability of the solution. The ammonio-citrate of iron has proved more reliable, and five grains of it in fifteen minims of distilled water, have been injected under the skin of patients with excellent results.

ADDISON'S DISEASE.

Rest is regarded by Dr. Pepper as the chief indication, and with it good hygienic influences and wholesome food. In some cases an exclusive milk diet has been found to do great good. The bowels and kidneys are carefully regulated. No long journeys are allowed. The system is sustained upon arsenic, iron, and cod-liver oil. Counter-irritation over the seat of the disease is useful in early stages, as is also faridization with mild currents. Nitrate of silver and iodide of potassium are administered upon general principles. The former drug has been found to be of great use where there is irritability of the stomach and intestines. Where the vomiting is violent and otherwise uncontrollable, chlorodyne is used. Where there is palpitation and dyspnœa, digitalis has been found to be invaluable. The forms of iron usually employed are the iodide and sesquichloride. A prescription frequently used is the following, recommended by Greenhow:

R Ferri susquichlo.
   Chloroformi... āā ℞ xv-xx.
   Glycerinæ..... .... f℥j.   M.
   SIG.—Three times a day, in water.

Not much confidence is placed in phosphorus. Alcohol is given in small quantities and in whatever form it is best assimilated by the stomach. Strychnia has done good in some cases.

CHRONIC PSORIASIS.

Dr. DaCosta uses an ointment of chrysophanic acid with the strength of one drachm of the acid to an ounce of simple cerate. This ointment is well rubbed into the skin of the whole body every evening. After this treatment has been persevered in for four days, the skin becomes much paler and smoother, and the scales begin to disappear.

PERITYPHLITIS.

In the treatment of this condition, the main indications are rest, alteratives and counter-irritation, in the shape of blisters over the site of pain, and induration. In some cases ice bags are of service as local applications. The bowels are kept open by mild laxatives. Indian hemp and bromide of potassium are often of service. The following drugs have been found to form an excellent laxative pill to be taken at bed time:

R Aloes.. gr. j.
   Ex. cannabis ind.. gr. ¼.
   Pulv. ferri sulph.............. .... gr. j. M.
   Et in pil. No. j. Divide.

If there is much hardening of the surrounding tissues a protracted course of the chloride of mercury and iodide of potassium in minute doses is prescribed by Dr. Pepper. In this condition pregnancy must be avoided.

CONSTRICTION OF THE AORTA WITH HYPERTROPHY OF THE HEART.

J. S., æt. nineteen, single, was admitted into the hospital on February 10th, 1879, and placed in Dr. James Hutchinson's ward. The patient's history shows that his mother had died of inflammation of the lungs, that his father is still alive, but suffers from heart disease, and that his brother and sister are in good health. In early life he appears to have been free from disease of all kinds, and positively states that he has never had rheumatism or venereal disease. His habits have always been temperate. About a year ago he had intermittent fever, while living near the Delaware, at Port Richmond, which was quotidian in type, and lasted, at intervals, throughout the winter. During the following spring and summer he seems to have been in his ordinary health, but in the fall caught a severe cold, which was accompanied by pain in the chest and small of the back, cough, bloody expectoration, and frequent micturition, which kept him in bed for two months. While convalescing from the acute symptoms of this attack, he suffered from dyspnœa, and noticed that his abdomen was beginning to enlarge. There has been no œdema of the lower extremities or of the face.

When admitted he was still suffering from cough, which was unaccompanied by expectoration of blood. The patient is anæmic, but not emaciated, and is rather small and undeveloped for his age. His appetite is good; his tongue moist and but slightly coated; his bowels are constipated and his hands somewhat cyanotic. He suffers from constant dyspnœa. His urine is passed in small quantity; it is of a clear amber color, and contains a slight amount of albumen ($1/12$ of bulk,) but no tube casts. Its specific gravity is 1025, and its reaction acid.

The impulse of the heart is felt in the sixth interspace external to the line of the nipple, and is heaving. Percussion shows that the organ is enlarged in every direction. At the apex a loud, blowing murmur is heard. At the base a systolic murmur is also heard, both to the right and left of the sternum, but less distinctly than at the point of impulse. These symptoms and signs must have been present for at least twelve months, as he states that he applied and was rejected by the medical examiner of the school ship about a year ago, on account of heart disease. There is a small amount of liquid in the peritoneal cavity, and the superficial veins upon the right side of the abdomen are very much enlarged, indicating probably some interference with the portal circulation.

Feb. 15*th.*—The effusion in the abdomen is increasing. The liver can now be plainly felt, extending three fingers' breadth below the arch of the ribs in the mammary line and also in the mid-sternal line. Its surface is smooth and free from nodules. The amount of urine passed in twenty-four hours is less than a pint. The patient has been taking since his admission Basham's mixture and broom tea, which have not yet, however, produced any increase in the flow of urine.

March 10*th.*—The patient has been steadily getting worse since date of last note. He constantly expectorates blood, and his dyspnœa is distressing to witness. All attempts to increase the urinary secretion have failed, although several different diuretics have been given. On the other hand, the abdominal effusion has been steadily increasing, until now his abdomen measures 38½ inches in circumference. With a view of diminishing the effusion he was ordered to have gr. ⅛ of elaterium.

March 16*th.*—Nothwithstanding the free purgation which was induced by elaterium, the effusion has increased to such an extent, and the dyspnœa is so distressing, that it was thought necessary to resort to tapping, when 10½ pints of liquid were removed. The patient bore the operation perfectly well. A day or two ago a teaspoonful of the fluid extract of jaborandi was given, but the depression which followed the sweating induced by it was so extreme that it was thought not wise to repeat the dose.

March 17*th.*—Patient died this morning at 7 o'clock. He reacted fairly after the operation and passed a comfortable day, but in the night was restless, and died in the morning apparently from syncope.

March 18*th.*—*Autopsy* twenty-seven and a half hours after death; chest and abdomen only opened. The lungs were found congested, especially the base of right lung. Throughout the left lung were numerous infarctions; the heart was very much enlarged, weighing when emptied of the fluid blood and clots which it contained, 24 oz. Its walls were in a relaxed condition, but were very much hypertrophied. All the cavities were dilated, especially the left ventricle. The mitral orifice readily permitted the passage of the four fingers of the hand, thus showing a marked insufficiency of the valve. The aortic semilunar valves were healthy, but just above the point of origin of the coronary artery, which was enormously enlarged, the aorta was constricted, the stricture only allowing the point of the little finger to pass through it. Beyond the stricture the aorta was small, being little more than half the usual diameter; the ductus arteriosus was closed. The liver weighed three pounds. Its projection beyond the arch of the ribs was not so marked as during life. External to the mammary line it was covered by them, but within the two mammary lines it extended fully two inches below them. Its substance was firm and congested. The gall bladder contained about half an ounce of thick, viscid bile. In the fissures of the liver were found some enlarged glands, two of which were situated so as to compress the portal vein at its point of entrance into the organ. There was thickening and œdema of the cellular tissue in the neighborhood, but no traces of previous inflammation. The spleen was not enlarged, but was deeply congested; the kidneys were also congested, but were otherwise apparently healthy.

THERMIC FEVER.

Dr. H. C. Wood advises that the patient should be immediately placed in an ice-water bath, and should be kept in the water just long enough to reduce the temperature, as taken in the mouth or rectum, to 100°, and no longer. After removal from the bath a hypodermic injection of quinia is given, to prevent another rise of temperature.

The headache, slight elevation of temperature, general distress and mental hebetude, which sometimes accompany convalescence from thermic fever, are generally found to yield to free blistering of the back of the head and neck, aided by small and repeated doses of mercury.

ACUTE TONSILLITIS.

Dr. Pepper's abortive treatment is by counter-irritation by iodine over the sub-maxillary region, and astringent washes to the tonsils, such as dilute Monsel's solution, or strong nitrate of silver (gr. lx. to f ℥ j,) solution. Guaiacum, in the shape of a three-grain lozenge, is given every three hours. In follicular tonsillitis, guaiacum has been found to effect a cure in the course of forty-eight hours in every case. The following prescription has been found to be of great service:

℞   Quiniæ sulph...gr. ij.
     Tinc. ferri chlor......................................♏ xv.
     Potassii chlor.. gr. v. M
Sig.—To be given every three hours, dissolved in syrup and water.

The patient is in all cases put to rest in bed and maintained on liquid diet. If the enlargement of the gland is very great it is lanced at once, provided pus has formed. As soon as the suppuration ceases, astringent poultices are applied to the neck.

COLLAPSE.

The hot bath is employed to restore caloric, and moderate doses of whiskey and ammonia are given internally. At the same time, from ten to fifteen minims of the tincture of digitalis are administered hypodermically, in order to restore cardiac action.

Rubbings with warm cloths dipped in the tincture of capsicum, followed by the application of dry heat and wrappings in blankets, in some cases replace the hot bath, but have not been found to be as efficient.—*Dr. Wood.*

ACUTE ANGINA.

Salicylic acid has acted most admirably in a case of very severe acute angina, attended with a great rise in temperature. Ten-grain doses of the acid, given at intervals of two or three hours, reduced the patient's temperature from 105° to 99° in the space of two days. This fall of temperature was permanent.—*Dr. Pepper.*

ACUTE AND CHRONIC RHEUMATISM.

Dr. Pepper continues to use salicylic acid as an antiphlogistic, in these cases, with very marked success. The dose in cases of acute rheumatism is ten grains every two or three hours, given either in powder, or diffused in mucilage of acacia. A very decided reduction of temperature is usually shown in several hours. In chronic rheumatism ten-grain doses are given thrice daily, with the like beneficial effects.

Dr. DaCosta prefers the salicylate of sodium to the pure salicylic acid.

Dr. Stille recommends in chronic rheumatism a prescription much used in England, the so-called "Chelsea Pensioner," from the fact of its being first used among the rheumatic old pensioners at the Chelsea Home.

Its ingredients are as follows:

R Sulphur .... ℥ij.
   Cream of tartar .... ℥j.
   Powd. rhubarb .... ℥ij.
   Guaiacum (resin) .... ℥j.
   Clarified honey .... ℔j.
   Powdered nutmeg .... ℥ij. M.

SIG.—Take two large teaspoonfuls at night and morning, for three days, in honey or mulled wine.

Dr. Pepper has great confidence in Zollikoffer's mixture:

R Pulv. resin. guiaci,
   Potas. iodidi .... āā gr. x.
   Tinct. colchici sem .... f℥ss.
   Aq. cinnam.
   Syrupi .... āā q. s. ad ℥j. M

SIG.—A dessertspoonful to a tablespoonful, thrice daily.

ACUTE AND CHRONIC DYSENTERY.

The patient is brought to the edge of a hard bed and placed in a position somewhat resembling that of lithotomy, his buttocks resting upon a hard pillow in such a way as to elevate the pelvis, and cause the injected fluid naturally to flow downwards and inwards. The "Simon gravity injection," is then employed in the following manner: a well-oiled, smooth, somewhat flexible, hard tube, with openings in the sides (an œsophageal tube is what Dr. H. C. Wood usually makes use of), and with a closed end, is gently and slowly introduced from eight to twelve inches into the rectum. The free, outer end of this tube is connected with a Davidson's syringe, and the fluid to be injected thus slowly pumped in. Sometimes the end of the œsophageal tube is united to a flexible India-rubber tube, in the end of which a funnel is placed. This funnel being then elevated some five or six feet above the patient's rectum, the medicated water is poured into it, and so by its own weight forces its way into the gut. Instead of using a funnel, the tube may be so arranged as to empty a bucket or other reservoir, which has been placed at the proper height.

The liquid injected is in all instances first heated to the temperature of the body, and so does not provoke peristaltic movement. The solution of nitrate of silver used by Dr. Wood, has a strength of ʒj to the three pints of water. The injection usually comes away in from five to ten minutes. A solution of common salt is always at hand, and should be at once thrown into the bowel if the nitrate of silver solution does not come away in the course of ten minutes. The same treatment has been also pursued with success in chronic dysentery.

Dr. Pepper uses a solution containing only ten grains to the quart, and introduces at first a pint and then a quart of the solution twice daily, at first and afterwards once in two days.

GOITRE.

Dr. Roland Curtin's successful mode of treatment has consisted in the hypodermic injection of from six to ten minims of a solution containing ninety-six grains of ergotina to the ounce of distilled water. The injection is repeated two or three times a week for the space of from four to six months, when the gland becomes thoroughly hardened. The gland begins to shrivel with the stoppage of the injections, and very soon returns to its normal size. Ergotina has been proved to be of no value in bronchocele, but only in cases of simple enlargement of the thyroid gland. The injection is attended with but very little pain, which pain is generally local, or referable to the origin of the sterno-cleido-mastoid muscle.

DYSPEPSIA.

Among drugs, arsenic in small doses, gradually increased, has been found to be a remedy of extreme importance. Where there is torpor of digestion joined with very marked sympathetic nervous disturbances, Dr. Pepper uses the following formulæ·

1. ℞   Sodæ bicarb.. ʒiij.
       Acid hydrocyan. dil........................gtt. xlviij.
       Tinct. valeriani.................................fʒj.
       Syrup zingiberis..............................fʒij.   M.
   SIG.—A teaspoonful thrice daily, in water.

2. ℞   Quiniæ sulph....................................gr. xvi.
       Strychniæ sulph................................gr. ¼.
       Acid. muriat. dil...............................fʒjss.
       Syrup. zingiberis...................q. s. ad fʒiv.   M.
   SIG.—Two teaspoonfuls in water, right after meals.

Where there is marked hepatic disturbance, the following prescriptions are excellent:

3. ℞   Acid. muriat. dil..............................fʒss.
       Tinc. nuc. vom.................................fʒss.
       Infu. gentianæ comp..............q. s. ad fʒiv. M.
   SIG.—A teaspoonful in water after meals.

4. ℞   Bismuth. subnit..................................ʒjss.
       Pepsinæ...ʒjss.
       Strychniæ sulph................................gr. j.
       Tinct. cardamon. comp............q. s. ad fʒiv.   M
   SIG.—A teaspoonful thrice daily, in water. If there is much flatulence, an increase is made in the quantity of bismuth and pepsin. If the case be merely one of gastric atony, the amount of strychnia is increased.

Where there is marked gastralgia, two to five drops of Fowler's solution are administered during the paroxysms. If this does not control the pain, a blister two inches square is applied to the epigastrium, and followed by a belladonna plaster six inches square.

Where the stomach is weak, its muscular action impaired, and its nerves over-sensitive, but little food should be taken into it at a time. The best diet is skimmed milk, half a pint every four hours. When milk is not well digested, lime-water is combined with it. Such foods as coffee, tea, and tobacco must, of course, be given up absolutely and at once. A sovereign article of diet is buttermilk. In buttermilk the cascin of milk is coagulated and broken up, so that the stomach is spared two steps of the regular process of digestion.

Another excellent preparation of milk is koumyss. It contains a good deal of carbonic acid. In all cases the stomach's work is made easier by a diet consisting of eggs, milk, starchy vegetables, stewed fruits, and a little butter, with stale bread.

DIABETES INSIPIDUS.

Dr. DaCosta puts the patient upon an initial dose of half a drachm of the fluid extract of ergot, thrice daily. This dose is gradually increased, first to 1 and then to 2 drachms. There is almost immediately a great reduction in the quantity of urine passed daily—from ten to three pints.

ULCER OF THE STOMACH.

Nitrate of silver, in the form of pills, is given in full doses half an hour before meals. If there is pain, opium, hydrocyanic acid, and chloroform are administered. An exclusive milk diet is the best. All solid foods are avoided. At the time of the hemorrhage, Dr. Pepper insists upon absolute rest. Pieces of cracked ice are swallowed by the patient. Monsel's solution, tannic, or gallic acid, are given internally. Morphia and ergotina are administered hypodermically, and all food for the time being (during the continuance of the hemorrhage) given by enemata.

POST-NASAL CATARRH.

Dr. J. Solis Cohen considers cleansing to be the principal factor in the treatment. The fluid which he employs for this purpose is tepid water at about blood heat, brought up to the specific gravity of blood, by means of some saline—such as the chloride or carbonate of sodium. The cleansing operation is persevered in until all the accumulations are removed. To cleanse the passages, the patient dips his nose into the vessel containing the saline liquid, and sniffs it up into the pharynx;

discharging it by the mouth; or the rubber-ball syringe, with a long curved nozzle, is used. Sometimes Thudicum's nasal douche is employed.

Another method of local treatment is by short flexible bougies, composed of gelatine impregnated with, for example, 2 grains of sulphate of zinc, and ½ grain of carbolic acid. A string runs through the bougie and extends some distance beyond it. This is for the purpose of securing the bougie so that it shall not drop into the throat. The bougie being inserted, the projecting portion of the string is fastened round the ear. The bougie generally melts in from twenty to thirty minutes.

LUMBAGO.

Dr. Pepper applies manipulation to the lumbar region of the spine, so as to restore mobility. To subdue the painful condition of the muscle, injections of $1/80$ of a grain of atropia and ⅛ of a grain of morphia, well diluted, are made well into the body of the muscle. Great care is had in the administration of morphia and atropia to nursing women, as belladonna is the most powerful anti-galactagogue known, and as too large doses of morphia not infrequently affect the child through its milk. The local application of blisters, iodine, and croton oil, together with the internal administration of iodide of potassium, are often serviceable.

LARYNGEAL PHTHISIS.

After enjoining absolute silence, Dr. Cohen applies the following solution to the affected parts by inhalation, or by the brush, douche, or spray, viz.:

℞ Acid. carb. concent... f℥j
    Tinct. iod. comp... ℥ij.
    Aquæ.............. ....q. s. ad f℥ iij M.

SIG.—One teaspoonful, applied three or four times daily in an ounce of water, by douche or spray, or thrown in boiling water, and the vapor inhaled. The advantages of the spray are, twofold—(1,) the water cleanses, and (2,) its impact upon the parts stimulates. The iodine acts as an absorbent, and the carbolic acid as a disinfectant and as an astringent.

When ulceration begins, more powerful applications are employed. Here the spray is still useful, for the part is cleansed at all times before any local application is made. Caution is needed, however, in making use of the spray, that no irritating substance is employed. The inclination of an ulcerating surface is not to heal; and an irritant, of course, throws a further obstacle in the way of healing. Where there are irregular excrescences on the vocal cords or other parts of the larynx, due to exuberant granulation following ulceration, they are, as a general rule, allowed to remain, unless they have grown to such a size as to interfere with respiration. If the arytenoid cartilage becomes ulcerated, the demand is for a soothing application, such as a weak solution of aconite and glycerine, or a weak solution of morphia and belladonna. The chloride of zinc, in the form of a weak solution, is an excellent application. The action of this remedy is limited to the points of ulceration. That astringent substance is always employed which does the least damage to the tissues surrounding the ulcers. The chloride of zinc is employed in a solution of one part to fifty, or even one hundred.

In the later stages of laryngeal phthisis the nitrate of silver does good; but it is used without the expectation of its curing. This substance forms a coating of coagulated albumen over the ulcerated tissue, and so protects it from the air and from the irritation of food. The solution of the nitrate of silver used has a strength of from forty to sixty grains to the ounce. It is applied merely to the ulcers at the margin of the larynx, and not to those in its interior, which it cannot benefit. Where the swelling of the parts interferes with respiration, and suffocation is threatened, tracheotomy may be necessary for the prolongation of life.

For the œdema Dr. Pepper uses sulphate of zinc, or copper, or alum. The following is an excellent gargle:

℞   Tinct. benzoici comp.. f℥ij.
    Glycerine.. f℥ss
    Aquæ... q. s. ad f℥iv. M.
Sig.—To be used as a gargle.

Inhalations of steam, of vapor of hops, or of conium, are employed as palliatives. Counter-irritation is applied externally to the larynx, in the shape of small blisters. To relieve the sense of fullness, lozenges of krameria, hæmatoxylon, or tannic acid are prescribed.

THE OPIUM HABIT.

Rest is the first item of the cure. The patient is kept in bed. Ample nourishment is given. If there is a morbid irritation of the mucus membrane of the stomach, the patient is placed upon an absolute diet of skimmed milk, beginning with a pint and running up to two quarts daily. All this time the quantity of opium taken is progressively diminished. If the conjoined milk and opium produce constipation, the condition is overcome by massage. The abdominal muscles are well kneaded daily. Regular injections of cod-liver oil are given in the morning, followed by cold water injections at night. Iron is administered in large and gradually-increasing doses. Dialyzed iron is the form of chalybeate generally employed. It is given in doses of from 10 drops to 1 fluid drachm thrice daily.—*Dr. Wm. Pepper.*

ANEURISM.

Dr. James Hutchinson has seen considerable benefit derived from the administration of very large doses of the iodide of potassium. In one case lately under his charge as much as 60 grains of this drug was taken thrice daily without producing *iodism*, and with the effect of quite markedly diminishing the size of the aneurism.

Another patient under the charge of Dr. Levis was put upon the iodide of potassium in gradually-increasing doses, beginning with 5 grains three times a day, and gradually increasing the dose until he finally took 3 *drachms* morning, noon and night, without producing iodism. He was also given aconite, as an adjuvant to the iodide of potassium, in much the same manner; beginning with 1 drop of the tincture of the root, and gradually increasing the amount to 5 drops three times a day. His diet, too, was regulated, and his manner of living changed. His diet was fair, as regards quantity, but was made up of substances which do not tend to fatten. Its composition consisted mainly of albuminous or nitrogenized articles of diet, with a careful avoidance of all kinds of stimulants. He lived as quietly as possible, lay down a great deal during the day, and kept himself free from any excitement. Rest is considered a very important element in the treatment of this class of cases.

Under this plan of treatment the man's condition improved very much. He has been slowly increasing in strength; there has been a marked improvement in his general health, and apparently the aneurismal sac is being gradually occluded. In other words, the aneurism is being cured by a consolidation of the sac.

Some time since, in the case of an aneurismal sac bulging from the subclavian artery, Dr. Levis carried pieces of horse hair through the walls of the sac, and so succeeded in producing partial coagulation of the blood.

INFLUENZA.

Dr. DaCosta's treatment is pre-eminently a supporting and expectant treatment, meeting special indications as they arise, and treating symptoms mainly. Good food (systematically given) and quinia are generally considered by him as of primary importance. Dover's powders or other diaphoretics are occasionally used. A certain amount of wine or milk-punch has been found to be indispensable, particularly in elderly patients. The amount of stimulus is proportioned to the severity of the attack.

ACUTE AND CHRONIC TUBAL NEPHRITIS.

The dropsy consequent upon this disease is treated by Dr. DaCosta with Basham's mixture and ♏xx of the fluid extract of jaborandi, three times a day. When the secretions of the skin and kidneys are fully re-established, the jaborandi is discontinued, and f℥ss of the fluid extract of ergot is given thrice daily in its place, the Basham being continued with the ergot. At the same time the region of the patient's kidneys is rubbed with croton oil. Exposure to cold and wet it avoided. Diet consists of milk and eggs and nourishing broths, with very little meat. Digitalis, belladonna and iron are often of service.

URÆMIA,

as a result of disease of the kidneys, is treated with the following prescription:

℞   Acid benzoici.. gr. xx.
      Syrupi tolutani.. f℥j.   M.
Sig.—To be given every third hour, largely diluted with water.

Benzoic acid has been found by Dr. DaCosta to prevent particularly the accumulation of the urinary salts in the blood. In addition to this treatment the patient is thoroughly purged with croton oil. He is also given a warm bath, or, if he is too weak, a vapor bath is taken in bed, by means of hot bricks wrapped in damp towels. In some cases dry cupping over the loins does good.

CHRONIC INTERSTITIAL NEPHRITIS.

Dr. DaCosta uses this formula:

℞   Tinct. ferri chloridi................................... ♏x.
      Syrupi limonis... ♏l.
      Aquæ... f℥iij. M
Sig.—To be taken thrice daily in a wineglassful of water.

A little wine is allowed and a nourishing diet given.

SURGICAL AND VENEREAL DISEASES.

FORIGN BODY IN THE URETHRA.

"On the 17th of July, 1878, T. D., æt. 63, was admitted into the University Hospital, for the purpose of having a fragment of a glass tube extracted from his urethra. D. gave the following history of his case: Has always been strictly temperate in his habits, and has never had any affection of his genito-urinary apparatus.

For several years patient has been one of a relief committee of a lodge of "Knights of Pythias," and while performing the duties of this position he met several members of this order who had great difficulty in passing urine, in consequence of enlargement of the prostate gland, and were dependent on the use of a catheter for relief. Fearing that he might have the same trouble at some future time, and being anxious to provide against such a disagreeable contingency, T. D. determined to practice catheterization on himself with a glass tube. In his first attempt he succeeded in passing a medium-sized glass tube into his bladder, and drew off urine, without the parts involved sustaining any injury. The success of this experiment encouraged him to try the operation a second time, the results of which were less satisfactory, as the sequel will show. On the day of T. D.'s admission into the hospital, he procured a straight glass tube, about 25 centimetres (10½ inches) in length, and half a centimetre in diameter, with which he proposed to repeat self-catheterization. He had no difficulty in introducing the tube as far as the prostatic urethra, but at this portion of the urethra the tube was suddenly arrested. D., in his efforts to overcome the obstruction used sufficient force to break the tube in two pieces, one of which he immediately pulled out of the urethra, but the other, being out of his reach, was left in.

Shortly after D.'s admission to the hospital, I made an examination, and found the fragment of the glass tube fixed in the urethra, one end in the prostatic portion, and the other penetrating the wall of the scrotal part. The fractured end of the fragment, D. had forced through the floor of the urethra into the loose cellular tissue of the scrotum, by his attempts to get the tube out. He had also tried to expel the tube by passing his urine; but instead of displacing it, it only served to carry urine into the scrotum, through the cellular tissue of which the urine was quickly diffused. From the position in which the foreign body lay in the urethra, and the manner in which it was held there, it was evident that it could not very easily be extracted by any instrument, such as forceps, passed into the urethra. Nor could it be removed through an opening made in the perineum without risk of doing additional violence to the already lacerated urethral walls. To take it out through a perineal section would necessitate the breaking of the fragment in two or more pieces in the urethra; the removal of which fragments would be attended with considerable danger, in consequence of the sharp, angular ends which the pieces would have. In considering the difficulties of the case from all points, it occurred to me that, in consequence of the fractured end of the broken tube being already in the scrotum, the safest and most practicable plan of removing the tube would be, either to make an opening in the integuments of the scrotum, and extract the tube through it, or to introduce my finger into the patient's rectum, get behind the smooth end of the tube in the prostatic urethra, if possible, and push the sharp end forward through the scrotal walls. As soon as the patient had been brought under the influence of ether, I directed two assistants to support his legs in the position that they would be held in operation of lithotomy, in order that the scrotum and perineum might be fully exposed.

I now proceeded to introduce my right forefinger into the

patient's rectum, carried it up behind the prostate gland, till I could distinctly feel the smooth end of the tube through the urethral wall; then, having made the scrotal integument tense near the fractured end, with the thumb and finger of my left hand, I quickly pushed the tube forward through the scrotal wall with my right forefinger. The small wound of the scrotum, made by the operation, was instantly closed by contraction of the dartos structure. As the scrotum had become infiltrated with urine prior to the extraction of the tube, I made four free incisions in it, two either side of the raphé, to relieve tension and to get rid of the urine, thereby to prevent subsequent sloughing. For eight days after the operation the patient's urine was drawn off three or four times a day with a catheter, the utmost care being used to prevent the escape of any urine through the wound in the urethra. At the expiration of this period he was permitted to pass his urine in the natural way; this he was able to do freely and without pain, clearly showing that the wound of the urethra had securely healed. Apparently, as a consequence of some injury inflicted on one of the testicles by the end of the tube, a mild epididymitis developed itself soon after the accident; this attack, however, yielded promptly to treatment, and soon subsided. The incised wounds of the scrotum that were made to facilitate the escape of urine from the cellular tissue, healed in a short time under the effects of local applications of laudanum and water. The patient's bowels were kept at rest for four or five days, and some slight pain relieved by suppositories of opium; no subsequent medication was required.

In the progress of this case it is worthy of notice that the patient did not have either a chill, or any symptoms of urethral fever. On the 30th of July, D. left the hospital quite well, with the exception of a little enlargement of one epididymis. The length of the fragment that was extracted was $10\frac{1}{2}$ centimetres (about $4\frac{1}{2}$ inches.") —*Dr. C. T. Hunter.*

FRACTURED PATELLA.

Dr. Thomas G. Morton has treated all the cases of fracture of the patella which have been brought into the wards for several years past with his improved modification of Malgaigne's hooks. In every instance the treatment has been permanently successful. Dr. Morton's first modification of the Malgaigne hooks consisted in making the hooks longer and straighter, but this improvement was shown to be of but slight advantage. Dr. R. J. Levis then made a further modification of the Malgaigne pattern, by cutting it into two separate pairs of hooks. This modification also was thrown aside after using it once. The last and most useful modification by Dr. Morton consists in cutting the hooks into two separate pairs, each pair consisting of one fixed and one movable hook.

At the University Hospital fractured patellæ are treated altogether by Dr. Agnew's splint. This consists of a flat posterior splint, with an eminence for the popliteal space, and with four rollers screwing in at the sides, two above and two below. Adhesive strips coming down on each side from above the broken bone are fastened to the lower rollers and screwed tight, and corresponding strips from below are secured in like manner to the two upper rollers. The fragments are thus securely brought together. The whole leg is then bandaged. This mode of treatment has given most excellent results.

Dr. Charles Hunter has lately invented a more simple apparatus for the treatment of these fractures, and has used it in one case very successfully. Extension is made by adhesive strips on each side of the leg, adherent from the groin to just above the seat of fracture. A weight is attached to the free ends of these strips, at the bottom of the bed. The whole leg is then tightly bandaged with figure-of-eight turns round the knee. This method will at once adjust itself to the necessities of country practitioners, by reason of its great simplicity.

HIP-JOINT DISEASE.

When patients cannot walk they are treated in the recumbent position. Dr. Agnew believes only in that form of treatment which fixes the diseased joint. He believes in allowing such patients to walk, if they are able to do so, but he does not believe in that kind of walking which causes either friction or motion of the affected joint. The immobility of the limb is secured by a splint, which may be applied either on the outer or the posterior aspect of the leg, thigh and body. The sound limb is raised by means of a cork-soled shoe, and, if sufficiently old, the patient is placed upon crutches. This plan has the advantage of fixing the diseased joint, and prevents the patient from resting upon the affected limb. Moreover, it admits of exercise in the open air. This treatment has aborted this disease in several cases in which it was begun early.

ELEPHANTIASIS ARABUM TREATED BY NERVE SECTION.

The patient was a colored man, 34 years of age, a farm laborer, and was first admitted to the Pennsylvania Hospital, on December 1st, 1873. He had been a slave in Virginia, where he was born and always resided until after the war, when he came to Philadelphia. His father was crippled by a fall, but was otherwise in good health; his mother was also healthy. He and a younger brother were the only ones out of fifteen children who suffered from enlargement of the limbs. He had never been in Barbadoes. Fourteen years before coming into the hospital he noticed the first symptoms of increasing size of the right leg. At first he had some pain, especially at night, but of late years he had been able to work as well as any one, and only felt inconvenience from the weight of the limb, and from a serous oozing that issued from some abrasions and kept his foot cold and wet. His general health was excellent, the disease being mainly confined to the right inferior extremity, which was markedly ichthyotic from the middle of the thigh down to the instep. The skin and superficial cellular tissue were very much hypertrophied, and hung in large folds over the ankle-joint. Under some portions of the thick, scaly epidermis there were effusions of pus. The left leg above the ankle was slightly enlarged, but the surface seemed natural. The skin of the abdomen showed impairment of nutrition and alteration of structure, which the patient said was the result of a scald received when he was young.

Dr. Thomas G. Morton tied the femoral artery at the usual place, on December 12th. The temperature of the limb, taken on the eighth day after the operation, was 98° between the toes, and $101\frac{1}{2}°$ on the surface of the calf, (the thermometer being held in position for one hour.) The axillary temperature was 99°. The limb was enveloped in a poultice of flaxseed, in order to remove the old epidermis. The ligature came away on the twenty-first day, and the small wound remaining speedily closed. The limb was then enveloped in a roller bandage firmly applied. This bandage was exchanged on January 7th, 1874, for one of India rubber, which was, however, removed at the end of four hours, on account of its producing numbness. The compression was, nevertheless, resumed as a regular part of the treatment, and the limb gradually and steadily decreased in size. The patient was discharged on March 21st, very much benefitted. [There was a marked improvement after the separation of the ligament, but, as firm compression was steadily maintained with the limb in a horizontal position, it was difficult to know how much of the improvement was really due to the operation.] The patient was subsequently admitted into the Philadelphia Hospital and, by the kindness of Dr. John H. Brinton, Dr. Morton had another opportunity of examining him. The limb was found to be nearly as large as it was prior to the ligation; the patient, however, considered himself much improved, and thought that the operation had markedly arrested the progress of the disease.

The man was re-admitted to the Pennsylvania Hospital, on November 9th, 1877. The right leg was found to be double its size when last seen, measuring *twenty-one* inches in circumference. From the exceedingly cumbrous character of the limb, the man desired to have an amputation performed. Having noticed the frequency with which operations for nerve section are followed by atrophy of the parts supplied by the nerve which is cut, Dr. Morton determined to attempt the artificial production of atrophy of the right lower extremity by section of the motor nerve of that limb. Accordingly, on November 17th, 1877, the *right sciatic nerve was laid bare, and one and one-half inches of its length exsected at the upper third of the thigh.* No unpleasant symptoms occurred after the operation incident to the section. There has been a steady diminution in the size of the limb ever since. On January 3d, 1878, the limb was measured, and found to be but *twelve and one-half inches in circumference, a reduction of some eight and one-half inches from the last measurement.*

An interesting feature of the case has been the desquamation of all the thick skin which covered the limb from the knee to the ankle and foot, especially about the lower third of the leg; patches of the skin, one-sixteenth of an inch thick, have peeled off from time to time, leaving a perfectly clear, soft, and pliable skin beneath. There has not been the least disposition on the part of the skin to ulcerate, and the lost sensibility is confined to the extreme anterior portion of the dorsum, all the sole of the foot, and a strip of the integument running directly up the posterior part of the leg to about the middle point between the heel and the popliteal space. This strip is about two inches in width on all portions of the leg, except this anæsthetic strip; the patient is able to distinguish between the compass points, provided they are held at not less than one inch apart. Later still the reduction to the size of the limb has become still more marked, so that when last seen the limb was almost reduced to its normal size.

GLEET.

"Where the system is very greatly run down, it is well to put the patient upon some such preparation as the citrate of iron and quinine in 10-grain doses, or 20 drops of the tincture of the chloride of iron, two or three times a day, with 2 grains of quinine. If, on the other hand, the patient is plethoric, he should be given some saline and an antimonial mixture. In treating local symptoms, it is useless to give the patient a syringe, and let him use it himself, for the injection employed must be applied directly to the diseased spots, if any benefit is to be derived from it, and no one can so apply it but the educated surgeon. Moreover, such a kind of application should be made with a special form of instrument, so that it may reach the diseased surface, and it only. There are a vast number of syringes, etc., which have been designed for this purpose. Some years ago I was accustomed to apply stramonium ointment and the nitrate of silver directly to the parts by means of one or other of these inventions. The application would, of course, cause some local irritation. The patient would be dieted and sent to bed, and some demulcent drink, such as flaxseed or slippery-elm tea, be administered. In other cases, again, I used the bulbed syringe, in which, after just passing the site of the stricture, the stream is projected backward. I have very often medicated strictures, by means of this instrument, with a strong solution of nitrate of silver, or with a mixture of cerate of opium and Goulard's extract. Of late months I have undergone a change of opinion in the matter of the treatment of gleet. I now think that there is nothing in the world so good as the introduction of nickel-plated conical bougies and the simple overstretching of the inflamed parts."—*Dr. S. W. Gross.*

CURVATURE OF THE SPINE.

In the suspension of a child for curvature of the spine, instead of using the head-sling, Dr. H. W. Wharton attaches a cord to the centre of the crosses over the top of the head of Barton's bandage. This simplifies the apparatus, and is less terrifying to the child.

ULCERS.

Dr. Charles Hunter uses the ozide of zinc as a protective in ulcers. This compound is smeared around the edges of an ulcer before applying the adhesive strips, and so prevents excoriation of the skin.

FRACTURED CLAVICLE.

Dr. D. Hayes Agnew treats fractured clavicle by perfect rest on the back in bed, with the head slightly elevated. But as the patient soon becomes restless, and as it is impossible to secure perfect quiet after the third day unless a nurse be secured to arrest every motion of the patient both by day and night, at the end of that time and when the ends of bone have, as it were, lost their disposition to get out of place, the patient is raised carefully from the prone into a sitting posture, and put in restraint by the introduction of an axillary pad four inches wide and five inches long, and tapering rapidly to a point, which elevates the arm and supports the shoulder, while a long strip of adhesive plaster, three and a half inches wide (Sayre's dressing), is passed round the body with a loop to support and elevate the arm.

SURGICAL SHOCK

is treated by Dr. Agnew by the hypodermic injection of carbonate of ammonium mixed with brandy.

CYSTITIS IN THE MALE.

If there is acute pain and spasm, blood is taken from the perinæum by means of leeches. The patient is kept in bed, and hot fomentations are applied over the pubes. At the same time he takes demulcent drinks with a little hyoscyamus. If the pain be violent, morphia is given by the mouth or rectum.

The following prescription is frequently employed:

℞ Pulv. opii... gr. xij.
   Camphoræ.. gr. xxx.
   Ext. belladonna... gr iij.
   Cacao... q. s.
      M. et in suppos. No. vi.; divide.
Sig.—One each night before retiring.

The diet employed is bland and unstimulating.

After the acute symptoms have abated, balsam of copaiva is given thrice daily in capsule, or rubbed up in an emulsion. To this, opium is occasionally added to prevent flatulence and griping. Turpentine in capsule is used with advantage in some cases. Other remedies are sometimes employed—such as buchu, decoction of trailing arbutus, or uva ursi with hops, highly recommended by Sir Henry Thompson—uvæ ursi, f ʒ iss.; lupulinæ, ʒ i; aquæ, Oij. This mixture is allowed to simmer for two hours; then enough water is poured in to bring the quantity of the mixture up to 2 quarts, and then 2 grains of morphia and 2 drachms of the bicarbonate of sodium are added. The dose of this mixture is a wineglassful.

In some cases benzoic acid has been found very useful. The initial dose is 5 grains, and the quantity is then gradually pushed up to 30 grains, and, in some instances, as much as 1 drachm is taken thrice daily.

A prescription very often employed by Dr. S. W. Gross and other members of the hospital staff is the following:

℞ Copaibæ balsami............. .... ʒ iv.
    Acid. benzoici.......,... ℈ iv.
    Gummi arabaci...
    Sacchari, ...āā ʒ ij.
    Gaultheriæ olei................ ....……...... gtt. xx.
    Aquæ camphoræ,........ ....……… q. s. ad fʒ viij. M.
Sig.—A tablespoonful every four hours.

As regards the treatment of the urine, the first thing that is done after dilating the stricture, if one exists, is to wash out the bladder. This is usually done in the following manner: A gum-elastic bag, holding about 4 ounces, is procured, and a basinful of water at 98° is made ready. Then, after drawing the urine with a flexible catheter, whose end is smooth, the bladder is washed out by simply connecting the mouth of the bag with the catheter, which has been allowed to remain *in situ*. The warm water thus injected is retained as long as possible, then drawn off, and the operation repeated.

If the urine is fœtid, 1 grain of the permanganate of potassium or a small part of carbolic acid is added to each part of warm water. When the water comes out clear, if it is desired to make an impression upon the mucous membrane of the bladder, a solution of borax (15 grains to the fluid ounce of water) is injected. At other times a 20-grain solution of the nitrate of silver is employed. If this solution does not cause too much smarting it is allowed to remain 10 seconds in the bladder.

The patient is kept at rest in bed during this treatment, and placed upon a very bland diet. (If there be any pain after the warm-water injections have come away, a little morphia is thrown into the bladder and allowed to remain there; or, if this does not afford relief, a hypodermic injection is given.)

In some bad cases it is necessary to perform cystotomy, or colotomy, in order to bring about a free discharge of fluid from the bladder.

Patients with cystitis are warned not to ride in wagons over rough stones; if possible, to refrain altogether from riding. They are advised to avoid all stimulating drinks, and to always wear flannel next their skin. They are not allowed to eat any greasy food, but are allowed plenty of fish and oysters. Their bowels are kept open by cold-water injections, or saline cathartics.

PROSTATORRHŒA WITH STRICTURE.

Where the stricture is fibrous, it is gradually dilated by means of Dr. S. W. Gross' expanding bougie. Where it is spasmodic and due to hyperæsthesia of the urethra, it is treated by the persevering introduction of conical steel bougies of gradually increasing sizes. If the meatus is unusually small it is slit up.

In obstinate cases of this nature astringent local applications are necessary, such as nitrate of silver, which is applied by means of the *porte caustique* devised by Prof. S. D. Gross. The solid nitrate is the form of silver used. After applying the caustic the patient is kept in bed, his diet reduced, and demulcent drinks administered freely.

If there be constipation, the bowels are kept well open by means of an occasional purge, or the patient takes as a routine medicine, each morning before breakfast, f ℥ vi of Hunyadi János water, or the following prescription, which was first originated by the late Dr. Robley Dunglison:

℞ Mag. sulphat... ℥j.
    Potass. bitart.. ℥j.
    Ferri sulphat.. gr. x.
    Aquæ.. q. s. ad Oij. M.
 SIG.—A wineglassful every morning before breakfast.

Another excellent plan which is often pursued is to inject five or six gobletfuls of cold water into the rectum each morning.

The diet provided is nutritious but bland. No coffee, tea, malt or alcoholic liquors, and no greasy or fried foods, are allowed.

For breakfast he is given grits or oatmeal, meat, and bread and butter; for dinner, plenty of good rare beefsteak or mutton-chop, with potatoes, tomatoes, etc. Supper should be a light meal.

If there be any anæmia, something like the following is given:

 ℞ Quiniæ sulph.. gr. xl.
   Tinct. ferri chloridi .... f℥ iv.
   Tinct. nucis vomicæ.. f℥ ivss.
   Aquæ..q. s. ad f℥ ijss. M.
 SIG.—A teaspoonful four times daily in water, through a siphon.

If, on the other hand, there is a tendency to plethora, depressants and purges are employed. The patient takes a sitz-bath every night and morning. He sleeps on a hard mattress, empties his bladder thoroughly before going to bed, and uses only the lightest of coverings.

To keep down the venereal appetite, 30 grains of the bromide of potassium are given thrice daily.

Where the condition is obstinate, local blistering is resorted to. A camel's-hair brush is dipped in cantharides and collodion, and a couple of lines are drawn with it, first on one side of the raphé of the perineum, and then on the other.

Intercourse is had regularly, so that the distended seminal vessels may be relieved, but it is never had at shorter intervals than every two or three weeks.

Where there is functional disease of the heart with palpitation, and flushing of the face, there is nothing better than digitalis in small and long-continued doses. It may be given in any of its medicinal forms.

NÆVUS MATERNA.

Two oiled pins are pushed right through the base of the growth so that they cross each other at right angles. A groove is then cut all the way round between the points of entrance and of exit of the pins with a sharp knife. A stout ligature is then passed round the base of the nævus and underneath the pins, and is drawn so tight as to completely strangulate the growth. Dr. S. W. Gross does not believe in temporizing in these cases by the use of the cautery or by the injection of irritating substances into the body of the tumor.

EXTERNAL HEMORRHOIDS,

are treated surgically by incising the pile with a bistoury, and then pressing out its contents, viz., the contained clot of blood. This slight operation relieves the pain and tension at once. As an after-treatment the parts are well bathed with cold water, and some medicine given to act on the liver and bowels. Dr. Samuel D. Gross does not at all believe in the immediate removal of these growths with the knife, as recommended by Erichsen and Bryant.

TAPPING IN DROPSY.

Dr. J. Solis Cohen treats obstinate cases of dropsy of the legs by tapping. The minute trocars which he uses are made of gold, with openings at the ends and on the sides. They have sharp three-cornered points. In inserting these canulæ they are thrust well into the flesh from below upward and then fastened in position by strings tied to their bulbous extremities. These strings are fixed by pieces of adhesive plaster. When the canulæ are fixed in position, slips of rubber tubing are fastened to their bulbous extremities. Where the fluid does not flow freely after the canulæ is inserted, its flow is started by pulling the rubber tubing between the fingers pressed tightly together, and so creating a vacuum. Care is had not to insert the canulæ too far down in the limb, where their presence may produce a sore, owing to the weight of flesh above them.

THE HYPERDISTENTION OF ABSCESSES.

This treatment of abscesses, first used by Mr. Callender, of London, has been tried with excellent effects at this hospital. The cases were both of them in the service of Dr. H. Lenox Hodge. In the first, that of a large femoral abscess in a child suffering with coxalgia, the abscess was first opened with a bistoury, and a large amount of pus discharged by gentle pressure. A 5 per cent. solution of carbolic acid in water was then injected through a common syringe, with tube attached, so as to completely fill up the abscess to overflowing. This hyperdistention caused a number of cheesy, shreddy, and fibrous particles to be discharged. No constitutional symptoms followed, and no pus was to be found subsequently upon the carbolized oil-dressing. The injection was repeated thirteen days later. At no time after the first injection were more than a few drops of pus exuded.

In the second case the results, though fair, were not so immediate and pronounced as in the first case; the temperature showing a constant tendency to rise and the pus still continuing to be discharged.

RAPID RECOVERY FROM DEPRESSED FRACTURE OF THE FRONTAL BONE.

The following case occurred in the service of Dr. H. Lenox Hodge: T. C., æt, 4, while playing on tow-path of a canal, was caught under his chin by the tow-line, and thrown violently down a bank estimated to be about twelve feet high. His head came in contact with a piece of cinder. This accident occurred at 6 P. M. on June 24th. The child was in the hospital by 8 P. M. Upon examination, a depressed fracture of the frontal bone about three-quarters of an inch above the superciliary ridge upon the outer edge of the frontal eminence, was found. The wound extended for some seven lines, and was accompanied by much laceration of the soft parts. The pupils were widely dilated. The breathing was stertorous and the condition that of general hebetude. The boy was not rendered insensible by the fall, but soon afterward lapsed into a comatose condition, only arousing himself to ask for water.

The treatment consisted in the coaptation of the sides of the wound with adhesive strips, and in placing over this a dressing of simple cerate. An ice-bag was applied to the head and a simple fever mixture given internally. The temperature immediately upon admission was 99° F. The next morning the boy was bright and cheerful; complained of no pain and asked for food. His pupils were contracted. As a precaution a catheter was passed, but its use was not again required.

June 26th.—Bowels not yet opened. A dose of castor oil given.

June 27th.—Bowels still inactive, but boy very bright and enters readily into conversation on any topic.

June 28th.—Bowels and bladder both acting regularly. Wound healing. Boy takes nap or two during day and sleeps regularly at night. Patient left hospital soon afterward, entirely well.

SERIOUS BURN FOLLOWED BY CONTRACTION OF TISSUE—CURED BY NITRATE OF SILVER AND PRESSURE CAREFULLY APPLIED BY MEANS OF ROLLERS.

J. T. F., æt. 10, admitted to the hospital on January 14th, 1879. This boy was badly burned, both above and below the elbow, and completely around the arm. Before coming to the hospital he had been treated for some six months at his home.

When first seen by the resident surgeon, Dr. J. M. Taylor, the burn was seen to have penetrated the muscles and showed pouting granulations of from two to three inches in diameter, encircling the arm. The elbow was much contracted.

When Dr. John Ashhurst, Jr., saw the case he thought that amputation would have to be performed. A dressing of oxide of zinc was immediately applied, the discharge of pus at that time being considerable. On the 20th of January, six days after admission, the granulations had very largely diminished under the use of the nitrate of silver, and in some places above the elbow the skin had united, forming narrow patches of healthy epidermis. On January 28th the granulations were all healing rapidly. On February 16th the arm was entirely cured.

The resident surgeon, Dr. Taylor, took great interest in the case, and each day straightened the arm out carefully as far as it would go, seeking to make it straighter at each visit. The nitrate of silver was applied with great thoroughness and care, and after each application the arm was bound up pretty tightly with turns of the roller over an oakum pad.

SPRAINS.

The routine treatment of sprains pursued at the Philadelphia Hospital, is that first introduced by the late Dr. Paul Goddard. The injured part is placed in a bath at the temperature of 70° F. The temperature of the water is then gradually raised until the extreme point of toleration is reached. This hot-water treatment has in every instance been followed by the most excellent effects.

GONORRHŒA IN THE MALE.

Dr. J. H. Brinton employs carbolic acid, in the shape of a solution containing gtt. ij of the acid to f ℥ i of lime-water, as a local application in the acute stage of the disease. At the same time cubebs is administered internally in doses of a tablespoonful in half a tumblerful of water three or four times a day. This dose is pushed until diarrhœa or nausea supervenes, when the quantity is reduced.

STRICTURE.

Dr. S. W. Gross considers that the only rational treatment of stricture consists in the restoration of the normal calibre of the urethra at the affected portion, and that the meatus should therefore be enlarged so as to admit of the passage of an instrument of the size adapted to restoring the constricted part (whose dimensions may be ascertained by means of the urethrometer devised by Dr. Gross) to its original dimensions. Dr. Gross claims, however, that *dilatation never effects a permanent cure,* and should therefore never be employed except in those cases where there is disease of other portions of the urinary apparatus.

He treats stricture almost entirely at present by internal urethrotomy, since he has determined from post-mortem examinations that there are very often sub-mucous bands which resist the action of the divulsor.

Having determined upon internal urethrotomy, he first allays the spasm and tenderness of the urethra by passing at first, and at long intervals, a moderate-sized conical steel bougie, gradually increasing the size of the instrument and decreasing the length of the intervals. If the meatus is the seat of stricture, or is smaller than the rest of the urethra, it is cut as a preliminary measure.

Ten grains of quinia are given before the operation, and the patient is made to pass his water so that the wound left by the urethrotome may have become glazed before the next act of micturition. The rectum is always emptied before the operation. Dr. Gross employs a urethrotome devised by himself, and made by Mr. G. Tiemann, of New York.

Immediately after the operation a conical steel bougie, whose size corresponds to that of the normal urethra, is passed and withdrawn, and $\frac{1}{3}$ of a grain of morphia is thrown under the skin. The patient is then confined to bed for forty-eight hours. The bougie should be introduced once every day for some time after the operation. Dr. Gross considers internal urethrotomy from behind forward the most effectual operation.

The treatment of stricture of the urethra followed by Dr. Brinton is not an exclusive one In the majority of cases he relies upon gradual dilatation, especially when the stricture is situated at or near the membranous portion of the urethra. In making dilatation he uses the ordinary Thompson's sound, and when much resistance is encountered he stretches the stricture rapidly with Thompson's dilating blades. In strictures of the spongy portion of the urethra, when, from any cause, slow dilatation is undesirable, he practices divulsion by Holtz's method, and also internal urethrotomy, using for the latter purpose an ordinary Charrière's instrument. He prefers urethrotomy to divulsion, particularly in irritable and resilient strictures. Strictures at or near the meatus he divides. In practicing the common process of gradual divulsion, when the stricture is tight and the opening small or tortuous, he usually employs whalebones bougies and Gouley's tunnelled catheters. The whalebone he makes himself, and they are very much finer than those to be bought in the shops, which, nine times out of ten, are not sufficiently flexible, and are therefore useless. Dr. Brinton insists with great emphasis upon the permeability of all strictures, with occasional exceptions, *i. e.*, those of traumatic origin. In difficult cases the passage of the whalebone is a work demanding great dexterity and gentleness, and, according to this surgeon's teachings, cannot be confidently anticipated if the patient has been practiced upon the same day with instruments of larger diameter and with rounded points. When a stricture is suspected, or known to be a tight one, he invariably uses first the whalebone and over it the tunnelled catheter, deferring to a later stage of dilatation, instruments flexible, or soft, of increased size.

In the practice of the Philadelphia Hospital, Dr. Brinton deprecates unnecessary interference with the urethra. The patients in this institution are paupers, collected from the lowest ranks of life, with constitutions utterly broken down by exposure and debauch. In his judgment the results obtained from operations upon such subjects are greatly inferior in point of success to those which attend like operations in private practice.

SYPHILITIC SORE THROAT.

The parts are kept thoroughly cleansed. This cleansing is performed with a syringe, brush or spray douche. The water used contains some of the chlorate or permanganate of potassium, or some carbolic acid. Local medication is not employed unless ulceration has set in. Occasionally a 20-grain solution of the nitrate of silver, or sulphate of copper is employed. In making these applications care is taken to cover the whole patch, so that the diseased tissue shall be completely destroyed.

Where Dr. Cohen desires to make a good local application, instead of a camel's-hair brush, he uses a broad or flat paint brush, so that one sweep of the brush will cover a space half an inch wide. In this way the whole diseased surface can be washed by one motion.

The best form of lunar caustic has been found to be that which comes in the shape of a lead pencil. This contrivance enables the physician to confine the cauterization strictly to the unhealthy tissue. When an application is made to the side of the palate, the wood is cut away from the pencil so as to leave a small piece of the caustic exposed laterally.

The constitutional treatment of the secondary syphilitic sore throat consists in mercurialization, and of the tertiary in iodization.

When perforation is threatened, the iodide of potassium is given in doses of from 30 to 90 grains every three or four hours, for thirty-six hours, if necessary, or until a change for the better takes place.

In giving large doses of the iodide, the patient's throat is carefully watched to see that œdema of the larynx does not occur.

If the patient comes in in such a late stage of perforation that the uvula is found to be suspended from its base by only a thin shred of flesh, Dr. Cohen's rule is to let it alone unless it gives rise to harassing cough, when it is clipped off with a pair of scissors. If this necessity does not occur, the strong probability is that the separated parts will unite again so soon as the system is thoroughly under the influence of the iodide of potassium.

With regard to the question as to how long the system should be kept under the influence of anti-syphilitic remedies, Dr. Cohen's rule is that these remedies should be continued until all evidence of the disease has ceased, and that they should even then be kept up for a couple of months longer, and then that small doses should be taken every few weeks, and whenever the throat shows the slightest disposition to take on specific inflammatory action.

TEMPORARILY IRREDUCIBLE HERNIA.

If the irreducibility is due to the distention of the sac by air, or by fæces, Dr. Agnew at once proceeds to attempt to dislodge the sac's contents. The patient is placed upon his back, his shoulders elevated, his thighs flexed upon the abdomen, and gentle compression instituted over the region of the tumor. This compression is made with great care and very gradually. If, at the end of fifteen minutes, a little yielding is felt and a slight gurgling sound heard, the prognosis is good. If this gentle compression is not followed by these signs it is stopped at once and something else tried. In the case of an inguinal hernia some leeches are placed over the course of the spermatic cord; if femoral, they are put above the saphenous opening, and a cold water dressing applied.

If the case is still obstinate, the patient is kept quiet on his back, and the following prescription given:

℞ Pulv. opii..gr. j.
   Ext. belladonnæ ....gr. ⁒
   Ext. aloës,
   Pulv. rhei.. āā gr. ij.
    M. et in pil. No. iv.; divide.
    Sig.—One pill every hour.

The cold water dressings are kept over the part. In the course of eight hours an injection is given. In cases where the stomach will retain anything, castor oil is given in doses of two teaspoonfuls every two or three hours, as a cathartic

[Dr. Chas. Hunter has performed Dr. Dowell's operation for hernia (strangulated) in a number of cases with positive success.]

A UNIQUE CASE OF INJURY TO THE BRAIN.

"E. A. D., æt. 40, on April 27th, 1879, walked into the hospital and up stairs. The top of his head was in a horrible condition. There was a lacerated wound three inches long by one and a half wide, and very deep. The integument had sloughed or been torn away, the bones were missing, the dura mater was exposed and sloughing, and the brain could be seen pulsating in the foul mass, which it seemed to be trying to pump out. The patient was perfectly rational, and gave the following account of himself:

He was first officer in a three-masted schooner sailing from the West Indies to New York with a cargo of sugar. He had worked very hard at loading the vessel in the heat, and for two days and nights after sailing for home he was constantly on the watch. He suddenly became ill, and conceived the idea that the captain and crew were going to murder him and throw him overboard. To frustrate them he thought he would take his own life, and tried to jump overboard, but was prevented from doing so. On the fourth day out, when off Cape Hatteras, he got possession of an axe and dealt himself several severe blows on the top of the head with the handle, fracturing the skull. He then, with the sharp edge, chopped

out the softened mass, and picked away pieces of the bones.
After this he got better and returned to normal consciousness.
On the 25th, four days after inflicting the injury, the vessel
arrived at New York. The patient walked to the cars and
came to Philadelphia.

He said he was a temperate man, that he only took an
occasional drink, and that he was not in delirium tremens.
His appearance certainly did not indicate a man of intemperate habits. His story was so unlikely that it was of course
doubted, but no inquiry elicited any other statement, and he
persisted in it to the last, bringing no accusation against any
one—in fact volunteering the declaration that the captain and
crew were the best he had ever sailed with.

The sloughs were partially cleared away, several small
pieces of bone were removed, cold compresses with mild antiseptics were applied, and a nourishing diet ordered.

The patient was in the third story. I feared he might
attempt to jump from the window, and so ordered him to a
private room in the basement. He remained there for one
night, but protested against it, and earnestly requested to be
sent back to his old quarters, saying that he knew perfectly
well what he was about, and that it was too much like a
prison down stairs. His request was complied with, and there
was no trouble.

By May 1st more serious symptoms set in. There was
high fever and a quick pulse. The temperature was $1\frac{1}{2}°$
higher on the right side than upon the left. This side was
the seat of spasmodic twitchings and contractions, and by the
4th it became completely paralyzed. A hernia cerebri was
also beginning in the wound, and there was some stupor.
There was no facial palsy. On the 5th there was an attack
of erysipelas, from which the patient recovered in a few days.
Discharges had in the meantime been thrown off freely from
the wound, and on the 7th an interesting event occurred. Dr.
McIlwaine, while dressing the wound, saw the end of a piece
of bone deep down in its anterior part and to the right side.

about an inch in front of the coronal suture. This and another piece he removed. These were pressing upon the right hemispheres in this position. There was some venous hemorrhage, and a small black clot, which was also taken out. The pieces were respectively ¼x½ an inch and ⅜x⅝ of an inch in size. They had been actually driven down and buried under the sound part of the skull by the force of the blows, and were only exposed by the cleaning up of the wound. Almost immediately the paraplegia in the left arm was relieved, and by the next day that in the leg had also disappeared.

The stupor was gone. The patient said he felt more rational, and he could bring his will to bear upon any desired movement, although he was weak.

This relief proved, however, to be but temporary. From day to day signs of brain disorganization and abscess were apparent. The paralysis returned, at first more manifest on the left side, but before death it became general. The intelligence was good up to within a few days of death, as rational answers were given to questions. There was no active delirium. There was much fever, the temperature chart showing a range from as low as 98½ on the 19th of May up to 106½ on the 31st. Control over the bladder and bowels was lost. On the 31st of May there was entire unconsciousness, and death took place at 2 P. M., June 1st.

Autopsy.—After the scalp was removed numerous linear chips of bone were found, and many linear scratches, twenty-five or more, which were made with a sharp instrument, were upon the outer faces of the frontal and parietal bones, and all were in a parallel line with the opening in the calvaria. This was 3 inches long and 1 inch wide. One inch in length was taken out of the frontal bone and the rest was from the parietals. There was an abscess leading from the base of the cerebral hernia to the corpus callosum. An abscess was in each hemisphere. Pus was infiltrated between the cerebrum and cerebellum. The dura mater was torn and thickened near the posterior portion of the wound, and here also there was abscess. The lateral ventricles were filled with serum. At the bottom of the longitudinal fissure there was a piece of bone ½ an inch long and 4 lines wide, about which there was an abscess.—*Dr. Wm. Hunt.*"

EXTENSION IN FRACTURES BELOW THE KNEE,

Dr. S. W. Gross treats in this way: The foot is well bandaged and covered with turns of a roller. A shingle is then cut to fit the shape of the sole and fastened to the foot by adhesive strips. The weight is attached to a knotted cord passed through the centre of this foot-piece. Potts' fracture is treated in this way, after first bringing the inverted foot into its normal position by means of a broad adhesive strip running from the inside of and across the middle of the sole, well up on the outside of the leg

FRACTURE OF THE CLAVICLE.

The arm is flexed and bound to the body. A silicate of sodium dressing is then applied so as to retain it in position. A pad in the axilla is necessary in lean subjects.—*Dr. J. H. Brinton.*

INTRACAPSULAR FRACTURES.

Instead of extension by adhesive strips and the use of sand bags to keep the fractured limb in position, Dr. S. W. Gross, by placing pillows under the knee, puts the leg into the shape of a double inclined plane. The adhesive strips are then attached on both sides, from the seat of the fracture to the knee, and are carried from that point straight out to a pulley and weight at the foot of the bed.

COMPRESSION OF THE BRAIN.

The treatment employed by Dr. R. J. Levis consists of rest in bed, with the head and shoulders slightly elevated, together with stimulating injections of turpentine and water. Turpentine beaten up with water (for it is not soluble in it) and thrown into the bowel with a syringe, in the form of a mixture, has been found to make a very good stimulating injection. It not only produces a stimulating effect upon the system, but causes what is equally desirable, a free evacuation of the bowels. In addition to this, an active mercurial cathartic is administered, and the patient is placed upon full doses of the bromide of potassium, 60 grains of this drug being given in the course of the first twelve hours, and subsequently reducing the dose to about one-half. In dealing with a patient in this condition, it is always borne in mind that if deglutition be impaired, as is usually the case, medicines and articles of food should not be given by the mouth, but rather by the rectum, because if a patient cannot swallow, there is great danger that whatever may be introduced into the mouth will pass down into the trachea and produce strangulation. For the same reason, if mucous accumulates in the trachea and bronchial tubes, the patient is inverted, to facilitate its discharge.

EPITHELIAL CANCER.

The treatment pursued by Dr. Richard J. Levis consists in the destruction of the cancerous tissue by successive applications of chromic acid. This acid destroys by rapid oxidation. The pure acid is prepared for application by diluting the crystals just enough to render them liquid, so as to permit of ready application with the brush. It is then applied to the margin of the cancerous growth, upon all sides, and the application is repeated from time to time, as the case may demand. In this way the morbid growth is gradually encroached upon

INGROWING TOE-NAIL.

Dr. Charles T. Hunter's plan of treating this condition, consists in introducing a thin layer of surgical cotton beneath the edge and extremity of the offending nail, thereby keeping the sharp edge of the nail separated from the ulcer, until the latter cicatrizes. A small roll, or pledget, of cotton is placed along the lateral margin of the nail so as to keep the prominent granulations pressed away from the nail. The dressing is kept in place by a film of collodion, and a narrow strip of adhesive plaster wound two or three times around the toe. Before the cotton is gently pressed, or crowded beneath the depressed nail-margin, collodion is painted over the surface of the ulcer, in order that the granulations may be compressed by the contractile film left by the evaporation of the ether contained in the collodion.

To introduce the cotton, Dr. Hunter uses a steel probe with one end hammered quite thin (about ¼ of a line in thickness,) and slightly curved on the flat—the curve corresponding to the lateral curve of the nail. The other extremity of the probe is likewise flattened, but not curved. The probe used is four inches long.

ERYSIPELAS.

Dr. Samuel D. Gross finds the best local wash to consist in a solution of the acetate of lead, (℥ss to Oij.) A cloth, saturated with this solution is placed over the parts and covered with oiled silk, or waxed paper. When the digestion is impaired calomel is administered, and followed in the course of six hours by a dose of castor oil. Quinia and the tincture of the chloride of iron are always administered for their tonic effect. If there is much circulatory disturbance, gtt. ij of the tincture of the root of aconite are given every four hours. Great attention is paid to the ventilation and diet. In the phlegmonous variety the best local treatment has been found by Dr. Gross to consist in free incisions. In every instance the patient is at once separated from the other patients, and if it be in summer he is quartered in a tent, outside of the main building, where there is plenty of fresh air and sunshine

PARAPHIMOSIS.

In mild cases Dr. S. D. Gross finds that the swelling can be reduced by the application of a solution of the acetate of lead, ($\frac{3}{3}$ ss of the acetate to O$\frac{3}{3}$ of water,) or by mercurial ointment. When the infiltration is great, the reduction of the swelling is secured in the following manner: The penis is grasped behind the retracted prepuce, and the blood is gradually forced out of the head of the penis, which is then pushed back, while the prepuce is pressed forward. When reduction is not to be accomplished by this compound movement, an incision is at once made through the stricture formed by the prepuce with a bistoury. In all cases where there is much infiltration, incisions are made into the infiltrated part, in order to allow the serum to escape. The penis, in this disease, is always so placed as to point towards the umbilicus.

THE MEDICAL AND SURGICAL DISEASES OF WOMEN.

PELVIC PERITONITIS AND CELLULITIS.

If the attack cannot be aborted, the treatment is taken up regularly, the two most important indications being (1) to stop the pain, and (2) to prevent the formation of pus. With these ends in view full doses of opium and of the bromide of potassium are given together with from thirty to forty grains of quinia daily. The abdomen is painted with iodine and covered with a poultice. If the woman is plethoric the morphia is given by the mouth with neutral mixture and wine of ipecac. In some cases tonics are demanded. Occasionally the application of belladonna and blue ointment locally proves beneficial.

If the attack lasts for more than a week and the local tenderness increases, the hot-water douche is applied to the tender cervix uteri. It is at this stage also that flying blisters are applied, beginning with a good-sized blister over the iliac region. In some cases this is all that is required. When the tumor still persists, however, another blister is put on over the womb, and then another over the other side of the abdomen, and so on until the swelling disappears entirely.

If at any time a sudden chill supervene, the plan followed is to begin all over again with large doses of quinia and of morphia.

When pus has formed tonics are administered, and among them iron especially.

In the later stages of the disease muriate of ammonia has been found to be a very excellent remedy.

The following prescription is that usually employed:

℞ Mist. glycyrrhizæ comp... .... ℔ ℥ vi.
    Ammonii chloridi.. ʒ ij.
    Hydrarg. chlo. corrosivi................................. gr. j.
    Tinct. aconiti radicis.................................... gtt. xxiv. M.
    Sig.—A tablespoonful in water every six hours.

If pus has formed, and it becomes impossible to secure its absorption by medicinal means, the spot is found where the abscess is beginning to point, and an incision made large enough to admit of a free drain of pus. After aspiration the cavity is injected with a solution containing one part of iodine to nine parts of water, or, in some instances, a five per cent. solution of carbolic acid is employed.—*Dr. Goodell.*

ANTEFLEXION OF THE UTERUS.

The patient was a servant-girl, 27 years of age, with a history of menstrual irregularities extending through a period of seven years. Accompanying the flow there had been occasional suprapubic pain. Six months before her admission to the hospital this pain had become constant, and she had been compelled to give up all work. She was obliged to pass her water twenty or thirty times a day. The urine was examined, and found to be entirely normal in all respects.

Vaginal examination showed the uterus to be a little lower than natural, the finger encountering the fundus in the anterior cul-de-sac. Together with this anteflexion there was some catarrh of the bladder, while the woman was anæmic and hysterical, and suffered greatly from constipation.

It was concluded by Dr. J. F. Meigs that the first thing to do was to build up the woman's general health. Rest in bed was enjoined; thrice a day she took gr. iv of the ammonio-citrate of iron with gentian, and the following prescription was employed, viz.:

℞ Magnesii sulphat .... ʒ vj.
    Acid sulph. dil.. ʒ ij.
    Ferri sulph... gr. xij.
    Quiniæ sulph.. gr. xij.
    Syrupi zingiberis... f℥ j.
    Aquæ... q. s. ad f℥ vi. M
    Sig.—A tablespoonful in ice-water thrice daily.

To cure the anteflexion, instead of introducing a pessary it was determined to persuade the woman to teach her bladder to hold gradually more and more urine. It was reasoned that when the bladder could hold twelve ounces the anteflexion would be largely reduced.

THE VOMITING OF PREGNANCY.

A good prescription is, viz.:

R Cerii oxalat.. .... gr. i.
    Ipecacuanhæ......................... .... gr. i.
    Creasoti.......................... .... gtt. ij. M.
Sig.—To be taken every hour.—*Dr. Goodell.*

This same prescription has been used with much profit at the Episcopal Hospital.

HÆMATOMA IN DOUGLAS' POUCH.

Dr. Goodell's plan is to keep the patient perfectly quiet and administer astringent drinks, such as sulphuric acid lemonade, etc. Opium enough is given to lull the pain and keep the patient thoroughly quiet. For a number of hours following the attack, but a very slight amount of nourishment is given. Stimulants are refrained from, and the patient is kept as low as possible, until all immediate danger from peritonitis has passed away. If the woman is married her vagina is packed with ice, but in the case of a virgin, where the hymen is intact, the ice is placed over the abdomen and perinæum.

PERIMETRITIS.

The first thing done by Dr. Goodell is to put the woman to bed and keep her quiet. Flying blisters are then applied over the abdomen. A mush poultice is put on as soon as the blister begins to draw. The first blister is placed on the right side of the abdomen, and then, in the course of a few days, another is applied to the left side, and then one over the middle; $1/24$ of a grain of the bichloride of mercury, with 10 grains of the muriate of ammonia, are given three times a day in the mixtura glycyrrhizæ composita. A pessary of cotton is constructed which can be so adjusted as to hold the womb up. This cotton is dipped in a solution containing ¼ of a grain of morphia to the drachm of glycerine. The morphia allays the pain and reduces the inflammation, and the glycerine usually sets up a copious watery discharge from the vagina. Iron is not employed until late in the progress of the disease.

After the inflammation is subdued, the patient is put upon the following mixture:

R Hydrarg. chlor. corros... gr. j.
   Liq. chloridi arsenitis... f℥ ss.
   Mist. ferri chloridi.
   Acid. muriat. dil...āā f℥ ij.
   Syrupi... f℥ iij.
   Aquæ ....q. s. ad f℥ vj. M.
Sig.—One tablespoonful after each meal.

POST-PARTUM HEMORRHAGE.

The management of post-partum hemorrhage is divided by Dr. R. A. F. Penrose into preventive and curative treatment. First, as regards preventive measures. If the woman is plethoric and has bled profusely at former labors, saline purgatives and diuretics are employed, and she is kept upon a low diet for some time previous to confinement. If there is marked anæmia, iron, bitters, stimulus, and plenty of good, nourishing food is used. If the labor proves long and tedious, it is hastened by the careful and skillful use of the forceps. If it is too rapid, on the other hand, the endeavor is made to render it slower by anæsthetics, etc. As soon as the child is born ergot is given freely, and all external means of causing contraction are brought to bear upon the case.

The curative treatment consists (1) in removing the placenta at once, in administering ergot, and in placing one hand in the cavity of the uterus and making firm pressure over the abdomen with the other. If these fail, (2) a cold application is at once made. A piece of ice about the size of a walnut, is carried right up into the cavity of the womb. If the ice does any good, it does it at once. If the patient still bleeds, (3,) a piece of rag is dipped into a cup of vinegar and then carried up into the uterus and squeezed, or a lemon is pared, gashed in numerous places, carried up into the womb and squeezed. The (4) last alternative consists in instituting compression upon the large vessels at the posterior part of the abdominal cavity, and in administering a good dose of opium.

When the hemorrhage is checked ergot is given freely, and a tight binding is placed over the lower part of the abdomen. A napkin is placed below the vulva to give notice of any return of hemorrhage. The patient's head is placed low down by removing pillow and bolster. If there is a marked disposition to faintness, the feet are elevated.

During convalescence the circulation is stimulated by laudanum, by the mouth, or morphia hypodermically. Alcohol is given, either in water or milk. The nerves are toned up by Hoffmann's anodyne, or by the sweet spirits of nitre. Plenty of saline food is given with milk. The rest of the treatment consists in watching the patient carefully and keeping her quiet.

At the Woman's Hospital, Dr. Anna E. Broomall controls post-partum hemorrhage by means of the hot-water douche. It is thus administered: the largest size of bed-pan is placed beneath the patient's hips. A Davidson's syringe is employed. The water used is brought to a temperature of 110° Fahr., and is not less in quantity than two quarts. A higher temperature is not well tolerated, and a lower temperature has been found to be ineffective. The amount of the injection has been given as not less than half a gallon, but the rule is to continue the injection until the return stream is clear.

Care is taken to see that the syringe is in good order, and that the air is thoroughly forced out of it before it is used. A metallic tube six inches in length is employed instead of the vaginal nozzle which comes with Davidson's syringe, and which is apt to act as a medium for puerperal infection. The metallic tube is inserted through the internal os up to the fundus of the womb. While the operator's right hand introduces the tube, the left hand grasps the fundus uteri through the abdominal parietes. The continuance and the frequency of the repetition of the douche depend on the promptness and permanence of the uterine contractions.

This same method of treatment has been employed with great success by Dr. Albert H. Smith.

ANÆMIA AND CHLOROSIS.

Basham's iron mixture, with the addition of fractional doses of strychnia, has been found very admirable in its effects. There are so many different recipes for making this celebrated mixture, that the one is here given which has proved to be the best:

℞ Tinct. ferri chloridi...f℥ iij.
    Acid. acetic. diluti...f℥ ss.
    Liquor. ammoniæ acetat.......................................f℥ iijss.
    Curacoæ,
    Syrupi simplicis.................................āā f℥j.
    Aquæ..q. s. ad f℥ viij. M.
Sig.—One tablespoonful after each meal.

The following formula makes another very elegant and generally useful preparation of iron:

℞ Tinct. ferri chloridi...f℥ ij.
    Acid. phosphorici diluti.......................................f℥ iij.
    Spts. limonis...f℥j.
    Syrupi simplicis..f℥ ijss.
    Aquæ..q. s. ad f℥ vj. M.
Sig.—One tablespoonful after each meal.

The dilute phosphoric acid is added both because it is a valuable nerve-tonic, and because it has the property of disguising the styptic taste of the iron, so much so that children readily take this mixture.

There are two other tonic preparations which are prescribed very frequently in the Hospital of the University of Pennsylvania, and with capital results. One of them is Blaud's pill, which Niemeyer extols so very highly:

 ℞ Pulv. ferri sulphat. exsiccat.
   Potass. carb. puræ.............................aa ℨij.
   Syrupi..q. s.
   Ut fiat massa dividenda in pilulas, No. xlviij.

During the first three days one pill is to be taken after each meal. On the fourth day four pills are taken during the day, on the fifth day five pills, on the sixth day six; that is to say, two pills after each meal. For three days more six pills are taken daily; then the dose is to be increased by one pill daily until three pills are taken after each meal. On this final dose the patient is kept for three or four weeks as the case may be. In stubborn cases the dose has been increased to the number of five pills thrice daily, and no other bad effects have followed it than a feeling of fullness in the head. This immunity is probably owing to the conversion of the iron sulphate into a carbonate.

The other preparation is a valuable alterative tonic:

 ℞ Hydrarg. chloridi corrosivi..................................gr. i.
   Liq. arsenici chloridi..fʒss.
   Tinct. ferri chloridi.
   Acid. hydrochlorici dil...........................aa fʒiv.
   Syrupi..fʒiij.
   Aquæ...q. s. ad fʒvj. M.
  Sig.—One dessertspoonful in a wineglassful of water after each meal.

Anæmic and chlorotic patients fatten and thrive wonderfully on this mixture. Dr. Goodell calls it the Mixture of Four Chlorides. It should not be given for a longer period than two weeks at a time.

where peritonitis is the most prominent element. In Bellevue Hospital five out of six are cured by this treatment. Besides the opium, his patients take a few doses of veratrum viride to diminish the frequency of pulse. He gives Norwood's mixture of veratrum in doses of gtt. v, when the opium has reduced respiration but not the pulse. It sometimes produces great nausea, attended by prostration and a tendency to syncope. Alcoholics are used when such effects are produced. He considers it a very good treatment to give opium and veratrum viride in alternate doses, and regards this as all that is necessary. In *metro-peritonitis* opium does not serve any important purpose, and he thinks it useless to give it, except to *soothe* the patient. Leeches to the vulva or perinæum and bleeding are very necessary. He very often employs injections of warm and tepid water into vagina and uterus. During the period of purulent infection, he prescribes quinia sulph., (gr. xv per day,) combined with morph. sulph., to reduce irritability. If there is a tendency to the formation of abscesses, food and stimulants are, of course, necessary.

VESICO-UTERINE FISTULA.

Dr. Bozeman's operation for the cure of this condition, consists in the complete excision of the anterior lip of the cervix uteri, together with a part of the vesico-vaginal septum, thus converting the original lesion into a vesico-utero vaginal fistula. This being done, the posterior lip of the opening, which was the stump of the cervix uteri, is next pared off, as is generally done in fistulæ of the latter class. Four silver-wire sutures are needed, which are introduced by a straight needle set in a curved needle-holder. This being done, the sutures are next adjusted in the usual manner, and a button or plate of lead of suitable form and size, is slid down upon them, and the whole then secured in place by the compression upon each wire of a perforated shot. The great utility claimed for this form of suture in the operation centres in the leaden

plate, which stands across the cervical canal and prevents its
recontraction and the consequent puckering of the line of coaptated edges, until union takes place. After washing out
the bladder, the patient is placed in bed and quiniæ sulph.
gr. x, and liq. opii comp. f ʒ j., administered per rectum, and
followed by gr. j of opium by the mouth, every six hours.
Dr. Bozeman claims great advantage in the operation from
the use of the self-sustaining and dilating speculum. By
means of it the vagina is expanded to the fullest extent, and
the greatest facility afforded to the movements of instruments.
The advantages of the knee-chest position are also fully illustrated. The patient, resting upon a supporting apparatus,
takes the anæsthetic with the greatest comfort.

nant. She may possibly be in the first or second week, but I consider it extremely unlikely. You may adopt the following general rule of diagnosis: When the cervix feels as hard as the tip of your nose, pregnancy does not exist; when it feels as soft as your lips, the womb probably contains a fœtus This woman has had no morning sickness, and her breasts, upon pressure, yield no milk, not even any moisture. Even early in pregnancy, as a general rule, the nipples will yield a drop or two of milk, when squeezed. Her bowels are regular, and she experiences no difficulty in voiding her urine, and on the other hand, no tendency to too frequent micturition. I think that the results of my examination preclude the possibility of pregnancy."—*Dr. Goodell.*

PUERPERAL FEVER.

The treatment pursued by Dr. Ellerslie Wallace is that of bleeding—*bleeding the fever out, no matter whether it require the loss of twelve or thirty-five ounces of blood.* When the fever is entirely subdued, the patient is given 3 grains of opium— of the watery extract of opium. It is then the custom to take a piece of flannel, broad enough to cover the distended belly, and after squeezing it out of hot water and rubbing some oil and laudanum well into it, to apply it to the distended surface, covering it with a bit of oiled silk or carded cotton, and pinning a bandage round the body loosely over all.

Together with the first dose of opium, the patient is given from 10 to 12 grains of calomel—this to re-establish the secretions. At the same time some nourishment is administered, in the shape of a little milk and some beef-tea, or a mouthful or two of some delicate broth.

When convalescence begins the amount of opium is diminished, and of broth increased, and the bowels are unloaded by means of a warm-water enema. This for puerperal peritonitis.

Puerperal metritis is treated by leeches, instead of venesection, together with opium and calomel.

Dr. Goodell treats this disease by intra-uterine injections of a warm 2 per cent. solution of carbolic acid. Ten-grain doses of quinia are given every four hours until marked cinchonism is produced. Morphia is administered in doses sufficiently large to relieve pain. The whole abdomen is painted with the compound tincture of iodine and covered with a large mush poultice. If it is deemed necessary to open the bowels, large doses of calomel are used.

DYSMENORRHŒA.

Where this condition is due to stenosis and flexion, Dr. Goodell does not believe in incising the cervix uteri, as some New York gynæcologists advise, but finds by placing the patient under ether and using very powerful dilators, whose blades do not feather, that it is very rarely needful to incise. In using these dilators, which have no shoulders, he first inserts the instrument right up to the fundus of the womb, and then withdraws it half an inch before beginning to dilate.

Another favorite method consists in taking some pieces of slippery elm bark, whittling them down to the size of matches, tying a string to each piece and packing the cervical canal with them. The only danger to be guarded against in using these slips, is that of pelvic peritonitis and cellulitis.

THE DIAGNOSIS OF OVARIAN CYST.

"Ovarian cyst is distinguished from dropsy in the following manner: In a case of ascites, the abdomen, when the patient is placed on her back, is flat on top and bulges out at the sides. Here there is a projection on top and not so much bulging out at the sides. In ascites, the intestines float up to the top, and we get resonance upon percussion. In this woman's case, both superficial and deep percussion reveal only flatness. In cases of ascites, when the fluid is allowed to settle, there is usually resonance on the top of the abdomen, and dense flatness at the sides. Here there is quite appreciable resonance at the sides. Examination of the external genitals, vagina, womb, and breasts, which have withered, excludes the possibility of pregnancy. There is one most certain way of settling the question, and that is by means of the aspirator. The fluid of ascites is straw-colored and limpid; that of a monocyst is perfectly clear and limpid, like spring water; that of a holycyst is thick, dark and turbid, from the presence of disintegrated red blood corpuscles; that of an oligocyst, which I suspect this to be, is usually of a milk-and-water, or of a light brown color. I should not think of tapping a polycyst unless I were ready to proceed at once to operate. The fluid is so intensely acrid and irritating that the escape of a few drops into the peritoneal cavity might set up a violent peritonitis, and rapidly destroy life."—*Dr. Goodell.*

LACERATION OF THE PERINÆUM.

Dr. Goodell advises the immediate operation which he has invariably found to be very successful in incomplete lacerations. In complete lacerations, however, it has not proved so successful as the secondary operation. In the primary operation, in order that the sutures may be most accurately put in, he recommends that ether be given, and that a sponge be placed high up in the vagina so as to stop the flow of the lochia, which embarrasses the operation. The stitches are made as in the secondary operation, and merely twisted together. In the secondary operation, if the sphincte ani is involved, he always imbeds the first two stitches. On the eighth day all the stitches are removed, except the lowest. The fæces are then softened by an injection of warm sweet oil, and the bowels moved twelve hours later by a dose of an ounce of castor oil, aided, if necessary, by an injection. After the bowels have been emptied the remaining stitch is removed.

Dr. Albert H. Smith uses the Baker-Brown needle for putting in the stitches. He always puts the first stitch in above, making that stitch draw thoroughly together the margins of sound tissue above the laceration. He believes in imbedding the wire all the way round in the tissues, so that when the ends of wire are drawn together there is no pocket left behind the stitches He always tightens the highest stitch first, thus protecting the tissues below from the flow of blood.

ABORTION.

" The treatment divides itself into two heads: the *preventive* and *curative* treatment. The preventive treatment is that by means of which we prevent the repetition of abortion. Here a knowledge of all the causes is indispensable. If plethora, anæmia, or nervous irritability exist, they should be modified or removed. If there is syphilis in either of the parents, they must undergo a prolonged anti-syphilitic and tonic treatment. This treatment must be steadily pursued for several years. Where the abortion is due to any local disorder, such as chronic endometritis, hypertrophy, prolapse, retroflexion, or erosion of the womb, the patient must be placed upon a steady course of treatment for the removal of these causes. Suppose, however, that we are called in to prevent the occurrence of a threatened abortion, what shall be done? If a syphilitic mother become pregnant, she must be subjected to a mercurial treatment and the local condition thus modified. If the ovum be already diseased, it will be impossible to prevent abortion. Indeed, under such circumstances, abortion would be a most fortunate circumstance, and on no condition should we attempt to balk nature in such a case.

The first question we should ask ourselves in all cases of threatened abortion, is: Why is it threatened? Upon the answer to this question depends all our treatment. If the ovum be diseased or dead, do not try to prevent the abortion. The family history will enable you to determine whether the ovum be diseased; if the product be dead, you will have the signs of death in utero. Suppose, however, that the product is not dead and not diseased, shall the abortion be prevented? The next question you must ask is: Can I prevent it? The embryo may be so far expelled that it would be worse than useless to interfere. The answer to this question depends on the dilatation of the os and the amount of hemorrhage. If the hemorrhage has been large, and the amount of blood lost considerable, the probability is that the utero-placental connections are so separated that abortion must ensue. Should a vaginal examination show the os uteri to be well dilated, and the membranes bulging, matters have gone so far that you cannot hope to prevent abortion, and therefore should assist it. There are various remedies for the hemorrhage. A drachm of the fluid extract of ergot may be given every three or four hours. I prefer the old wine of ergot given in doses of 2 drachms every hour or so. In addition to the ergot, the books tell you to apply cold cloths, ice, and vinegar to the abdomen, and to give gallic and tannic acid, or acetate of lead, internally. I don't believe in any of these remedies. They are not-half direct enough for me. If ergot does not control the bleeding, the next best thing you can do will be to tampon the vagina. A tampon acts in two ways: (1) it plugs the vagina; and (2) it stimulates the uterus to increased efforts. Never, however, put in a tampon unless you have given up all hopes of preventing the abortion. If you don't happen to have sponge-tents, tear up any napkin or piece of cloth that comes to hand and plug up the vagina. Sometimes I have had most excellent results from tamponing the mouth of the uterus with a sponge-tent or laminaria.

Now, as regards the curative treatment. If abortion occur during the second or third month, strive to secure a complete evacuation of the uterus. Otherwise the placenta will remain behind, and, becoming detached in the course of a day or so, will give rise to very serious hemorrhage. If the abortion occur in the fourth, fifth, sixth or seventh months, the membranes may be ruptured without danger if the hemorrhage proves excessive. When the child has been expelled, introduce your finger into the cavity of the uterus, and feel for the afterbirth. If it still remains in the cavity, it must be removed at all hazards. Have the woman brought to the edge of the bed, place one hand upon the abdomen, and insert a finger of the other hand into the uterus. Press the uterus well down upon your finger inserted into it, and scrape away until you have removed all the afterbirth. Never leave this work to nature, but see to it yourself at once. Where the placenta and membranes cannot be removed by the finger, various instruments have been devised for seizing them and bringing them out. Hodge's modification of Everett's bullet forceps is an entire failure, for the simple reason that you can't hold on to any foreign body after you have seized it with this instrument. Dewees' hook is not open to this objection, but it is so sharp that it might do much injury. The only instrument that has proved efficient is this species of duck-bill forceps which I now show you. I defy anything to get away which has once been caught by this instrument. If the placenta and membranes have begun to putrefy when you remove them, wash out the cavity thoroughly with some antiseptic injection.

Suppose that there is a hope of preventing abortion, what must you do? Put the woman to bed and give her 2 grains of opium by the rectum. Or else you may use a rectal suppository containing $\frac{1}{2}$ a grain of the extract of belladonna, and 1 grain of the watery extract of opium. Opium may be given hypodermically where the case is an urgent one. The bromides, too, should be freely administered in doses of from 25 grains up to a drachm. When the symptoms are very acute, ether or chloroform may be inhaled. Occasionally, where the woman is plethoric, from six to eight ounces of blood should be taken from the groin by leeches, or from the arm by venesection. Dry cups or mustard plasters may also be applied to the sacrum.

When abortion takes place the symptoms will generally disappear. There willl be, perhaps, a lochial discharge for a day or two. You must now employ the after-treatment for labor at term. Keep the woman in bed for two weeks or more. Regulate her bowels by mild saline laxatives; do not use enemata at this time. Be careful, also, not to feed the patient too highly. Do not allow any meat for the first week. Ergot should be given steadily to promote the proper subinvolution of the uterus. When the patient gets up, put her on a tonic treatment, and impress it upon her that she is still an invalid, and must take no violent exercise. At her next menstrual period the flow will probably be excessive. This should be treated by rest and by saline laxations. The vagina should be well cleansed with astringent solutions, and ergot, gallic acid, and cinnamon tea administered."—*Dr. R. A. F. Penrose.*

TREATMENT OF THE FUNIS.

As soon as the child cries lustily Dr. Goodell cuts the cord, and the umbilical portion being firmly held by the thumb and forefinger, the free end is "stripped" of Wharton's jelly and of any blood that may remain in it. Any blisters of Wharton's jelly which remain unemptied after this "stripping," are nicked and squeezed out. After the removal of the pressure of the thumb and forefinger, all bleeding usually ceases, and then the cord is tied. No subsequent dressing is thereafter used, for the cord rapidly dries up without smell, and drops off without leaving a sore behind.

GATHERED BREASTS.

When an abscess forms in the mammary glands, Dr. Goodell practices an early incision. If the pus lies deep, or is lodged behind the gland, a cutaneous incision is first made, a grooved director is pushed into the abscess and the opening is enlarged by the uterine dilator. The breast is then tightly strapped with adhesive plaster and treated by a dry compress of oakum. Should the abscess show symptoms of becoming chronic, its walls are overstretched by an injection of a 3 per cent. solution of carbolic acid. This overdistension is practiced in order that the acid may reach every nook and cranny of the purulent cavity.

OVARIOTOMY.

Dr. Goodell invariably adopts the following procedure: A 5 per cent. solution of carbolic acid is used in the spray, and all the instruments and sponges are immersed in a solution of the same strength. The pedicle is treated by the intra-peritoneal method, being transfixed, tied and dropped within the abdominal cavity. The peritoneum is always included in the stitches which close the abdominal wound. All obstinate bleeding points are tied with a gut ligature, but the pedicle itself is secured by fine carbolized silk. In three cases where there were numerous adhesions the glass drainage-tube was employed. The dressing consists merely of salicylated cotton, held in place by adhesive straps, the whole secured by an elastic flannel binder. Dr. Goodell prefers this dry form of dressing to Lister's wet dressing. The after-treatment consists of opium enough to allay pain, and in one tablespoonful of milk combined with lime-water, given every two hours for the first forty-eight hours. As soon as wind escapes from the bowels, the supply of food is increased.

[The patients are prepared for the operation by a soap-bath on the previous evening, and by the administration of 1 grain of opium at bedtime. Dr. Goodell always operates at the hour of 11 A. M., as being the time at which the vital forces are at their best.] When high temperature ensues, it is reduced by the ice-cap, which has been found to act very efficiently.

SORE NIPPLES.

Chapped nipples are treated either by a 20-grain to 1 drachm to the ounce solution of the glycerole of the nitrate of lead, or by a mixture of 2 drachms of iodoform to the ounce of the balsam of Peru. The balsam is used because it disguises the smell of the iodoform.—*Dr. Goodell.*

PLACENTA PRÆVIA.

Abnormal discharges of blood from the womb as early as the sixth month of pregnancy, are treated by putting the patient to bed, keeping her quiet and administering from gr. ½ to gr. ¾ of opium, and gr. ij to iij of sugar of lead in f ℥ ss. of the infusion of rose. But slight trouble is experienced in controlling hemorrhage until the neck of the womb begins to dilate in earnest, and in so doing tears the placenta loose from the uterine sinuses. When this occurs the only effective treatment has been found by Dr. Wallace to consist in the immediate tamponing of the vagina with sponge tents. Cathartics are not thought of, even if the rectum be overloaded. The tampon is not on any account removed until it begins to protrude from the vagina, and when it cannot be pushed back, though much force be expended in the effort. The minute the first sponges are taken out their places are supplied by fresh ones. When, in due course of time, the os uteri is found to have expanded sufficiently, the hand is greased well, passed far up into the uterus—forearm and wrist acting as tampons—as much of the placenta is separated as may be necessary, and the child is turned and delivered with the greatest possible speed. The moment the child is born the placenta is removed by the hand, and the uterus is made to contract completely by means of ice, pressure, ergot, etc., and a bandage is firmly applied round the patient's abdomen.

THE DIAGNOSIS OF CONGENITAL SYPHILIS.

Dr. Goodell says that congenital syphilis may appear either before or after birth. The labor is usually premature, and the first symptom of the disease is the hoarse cry to which the child gives utterance. The bullæ soon show themselves. The disease in uteri takes the form of placentitis, the exudation presses the blood out of the small capillaries and so gradually starves the production of conception, or there may be a gummy tumor or a fatty degeneration of the placenta, so causing premature labor. In some cases the labor is precipitated by atheroma of the vessels of the cord.

The child gives utterance to the hoarse cry because there are already syphilitic ulcers on the mucous membrane of its throat and air-tubes. Such children are always puny and sickly looking. The bullæ appear in the course of a few hours after the birth, and are visible on either the scrotum, hands, or feet.

If the disease does not reveal itself at birth, it usually appears some time between the second week and third month after birth. The first symptom is excessive crying of the child at night. This crying is caused by the incipient bone disease—pains in the bones. Its cry, too, is muffled and hoarse. The next symptom is the snuffles; the child's nose is all stopped up—a scalding coryza comes on. Then the child grows wizened and thin, and its skin lies all in rolls and wrinkles, and is more like parchment in consistency than skin. The so-called *copper* maculæ show themselves, or the complexion gradually assumes a coffee-and-milk hue. Then the eruption comes out all over the body, and stamps the case indisputably as one of syphilis.

In some instances it is very hard to find out whether the mother has the disease herself, or whether the father has been the only instrument of innoculation. In this connection the curious fact has often been noted that an apparently unaffected mother can nurse her syphilitic child with impunity, whereas the child is sure to communicate the disease to a healthy wet-nurse. Should a syphilitic child be knowingly allowed to draw its sustenance from a wet-nurse, and the nurse to be so innoculated, she has just grounds for an action at law against the parents of the child.

Treatment is generally hopeless. It is a wise law of nature which sentences all such vitiated and diseased products to early death. Of course all the physician can do is to subject the child to a brisk mercurial treatment.

HABITUAL CONSTIPATION IN THE FEMALE, WITH FISSURES OF THE RECTUM.

Dr. Goodell cures fissures of the rectum by over-stretching the sphincter ani. To do this he inserts his two thumbs into the rectum and then pulls them apart until either the sphincter begins to yield, or he can feel the rami of the ischia on each side. To do this requires the exercise of considerable strength.

As regards after-treatment, the patient is taught to go to stool regularly every day, and to eat certain kinds of food only. For medicine the following prescription is given:

℞ Ext. colocynth. comp...gr. ij.
    Pulv. rhei.........   ....gr. j.
    Ext. belladonnæ..gr. ¼
    Ext. hyoscyami..gr. ss. M
         Et in pil., No. 1; divide.
    Sig.—To be taken at bed-time.

In some cases $1/20$ of a grain of strychnia is added to the above with advantage.

AMENORRHŒA.

In amenorrhœa from anæmia and chlorosis, Blaud's pills are prescribed. (See *Anæmia and Chlorosis*.)

To counteract the possible costive effect of the sulphate of iron in these pills, Dr. Goodell advises the following aperient mixture: the pulvis glycyrrhizæ compositus of the Prussian Pharmacopœia, upon which he keeps patients for months and always with benefit.

℞ Pulv. glycyrrh. rad.
Pulv. sennæ..āā ℥ss.
Sulphuris sublim.
Pulv. fœniculi..āā ʒij.
Sacchar. purif..℥jss. M.
SIG.—One teaspoonful in half a cupful of water at bed-time.

Or the following :

℞ Ext. colocynth comp................................gr. ij.
Ext. belladonnæ.......................................gr. ⅓
Ext. gentianæ..gr. j.
Ol. carui..gtt. ss. M.
Et ft. pil., No. j.
SIG.—To be taken at bed-time.

Where the disease is due to torpidity of the ovaries this prescription is used by Dr. Penrose:

℞ Ex. aloës..ʒj.
Ferri sulphat. exsic..................................ʒij.
Assafœtidæ..ʒiv. M.
SIG.—One pill after each meal.

This number to be gradually increased to 2 and then to 3 pills after each meal. If the bowels are at any time over-affected a return is made to the initial dose.

Another excellent prescription used by Dr. Goodell is the following :

℞ Tr. ergotæ...f℥ij.
Decoct. aloës comp................................q. s. ad f℥ viij. M.
SIG.—Two tablespoonfuls, twice daily.

CARUNCLE OF THE URETHRA.

In timid women, who refuse to submit to an operation, Dr. Goodell either mummifies the growth with crystallized carbolic acid melted down by heat, or destroys it by applications of chromic acid, made with the utmost care by means of a match whittled down to a point, the excess of acid being afterwards neutralized by injections of a strong solution of the bicarbonate of sodium. If an operation is permitted, he cuts off the growth with a pair of scissors curved on the flat, and sears the wound with a hot wire, or with *Paquelin's Thermo-Cautere.* He advises the use of an alcohol lamp for heating the wire, because when an ordinary light is used the impression upon the operator's retina, made by the bright flame, so obscures his vision that the cauterizing apparatus grows cool before he can clearly see the point where the application is to be made. He hastens the healing of the cauterized surface by occasional applications of carbolic acid, or by dusting it with iodoform.

CANCER OF THE CERVIX UTERI.

Whenever practicable the whole cervix is removed, by either the hot or the cold wire. If this cannot be done, Dr. Goodell removes the malignant growth by scraping or by the gouge forceps, and the surface is then charred by the thermocautery. This radical treatment is reinforced by subsequent applications of the ethylate of sodium. In these operations upon the cervix he finds injections of ordinary vinegar to be an excellent means of controlling any embarrassing bleeding. By these means he has succeeded in curing several cases.

CYSTITIS IN THE FEMALE.

Transient cystitis, dependent upon obscure causes, Dr. Goodell treats by rectal suppositories, containing 1 grain each of the aqueous extract of opium, and of the extract of belladonna. Hysterical cases generally yield to *massage* and electricity. In obstinate cases of bladder trouble Dr. Goodell warmly advocates the dilatation of the urethra throughout its whole length by the introduction of the forefinger. In the therapeutical treatment of this troublesome disorder, atropia has been found to be the most efficient remedy. It may be combined with alkalies or acids, according to the condition of the urine. Injections of a 2-grain solution of quinia into the bladder, together with large doses of the same drug by the mouth, will often improve the condition of the patient. In very bad cases the most efficient injection, perhaps, is one of the nitrate of silver, beginning with a weak solution and increasing its strength daily, until 30 grains to the ounce is reached. These strong solutions should not remain in the bladder longer than ten or fifteen seconds. All malpositions of the womb must of course be rectified, especially if they have any bearing upon the disease.

CONICAL CERVIX UTERI.

This condition is treated either by forcible dilitation with a strong uterine dilator, or by lateral section with a hysterotome. If the cervix is sickle-shaped, Dr. Goodell performs the section of the posterior lip. The subsequent treatment consists in such local management as tends to keep the parts from closing up.

PROLAPSE OF THE OVARIES

is best managed by the knee-breast posture, and by the administration of such alteratives as tend to lessen the congestion of these organs. Among these, Dr. Goodell deems the best to be the chloride of ammonium in combination with the bichloride of mercury. Sometimes, however, large doses of the bromides act very happily. He considers this dislocation to be due in a great measure to the congestion of the sexual organs, consequent upon the use of measures to prevent conception, or from masturbation. He finds that pessaries are rarely useful in this distressing condition. But among them he thinks Cutter's bulb pessary to be the best. He considers prolapsed ovaries to be a very frequently overlooked cause of many pelvic aches and pains, which are attributed very generally to the womb alone.

CLOSURE OF THE VULVA FOR VESICO-VAGINAL FISTULA.

In curable cases of vesico-vaginal fistula, in which the urethra has been destroyed, Dr. Goodell has twice succeeded in wholly relieving the patient. In one case this was done, by making an artificial recto-vaginal fistula, and in the other by leaving an already vesico-vaginal fistula intact, and in then closing up the vulva. Whenever practicable he prefers in this unfortunate condition, provided the urethra is unimpaired, to close up the vagina as high up as possible, so that the marital relations should not be interfered with. In uncomplicated cases of vesico-vaginal fistula he prefers the use of shot to twisting of the wires, because they form permanent adjusters and prevent eversion of the edges of the wound.

VAGINITIS.

Non-specific and acute cases of vaginitis Dr. Goodell treats by such hot and emollient injections as flaxseed, or slippery elm bark tea. When the inflammation has subsided vaginal suppositories, containing 5 grains of iodoform, are ordered twice or thrice daily. In the chronic forms of this complaint he uses suppositories of tannin or iodoform, or long tampons of absorbent cotton dipped in astringent solutions of acetate of lead, and of zinc to which laudanum has been added.

FUNGOUS VEGETATIONS OF THE ENDOMETRIUM.

In this condition Dr. Goodell removes the unhealthy growth, either by the dull or by the sharp curette, or if the os is sufficiently patulous, by means of a small fenestrated polypus forceps. The uterine cavity is then cleansed with a saturated tincture of iodine if the cervical canal is not very open, but when it is gaping he prefers Monsel's solution. In the former condition he avoids the use of the iron, because it forms clots which cannot easily be expelled from the womb without causing much pain. He deems the iron, however, the more efficacious treatment of the two.

VAGINISMUS.

Dr. Goodell had never yet been compelled to resort to the deep posterior incision practiced by Dr. Marion Sims, although in two cases he was obliged to snip off irritable carunculæ myrtiformes. He treats this disease precisely as he would treat an anal fissure. If the local spasm does not yield to constitutional treatment and to vaginal suppositories of morphia, belladonna, carbolic acid and iodoform, he puts the woman under ether and forcibly stretches the vulvo-vaginal opening either by means of the two thumbs, or by the fore and middle fingers of each hand.

NERVOUS DISEASES.

WARDS IN THE PHILADELPHIA HOSPITAL FOR DISEASES OF THE NERVOUS SYSTEM.

The cases found in these wards, which are in charge of Dr. Charles K. Mills, neurologist to the hospital, are chiefly examples of chronic organic disease of the nervous system—hemiplegies from hemorrhage, thrombosis, or embolism; cerebral, cerebro-spinal, and spinal scleroses; meningitis, meningo-encephalitis, and meningo-myelitis; epilepsy, hystero-epilepsy, and hysteria; brain tumors, spinal softening, and the like. Acute cerebral and spinal disorders; neuralgias, peripheral paralyses, local spasmodic diseases, and similar affections, are sometimes, but not so frequently, represented.

ELECTRICITY.

In connection with the wards, a large apartment, known as the *Electrical Room*, has been fitted up. It contains one of Flemming & Talbot's permanent batteries of sixty cells, and a fine faradic instrument from the same manufacturers. The wards are also supplied with portable galvanic and faradic instruments.

Dr. Mills, during the past year, has used electricity with marked success in the treatment of bed-sores, which, in spite of the best of care, are apt to form in cases of spinal and cerebral disease. The "silver-and-zinc-plate" method is the one generally employed, a silver plate being placed over the sore, and a zinc plate (connected by a wire with the silver) on a piece of acidulated chamois skin or paper lint, which rests on the unbroken skin a few inches above. A weak current from the galvanic battery is sometimes used instead of the plates. A silver plate applied to the sore is connected with the negative electrode; an ordinary rheophore, joined to the positive pole, being placed upon the surface near. The *seance* is continued for from five to ten minutes daily. Many cases of chronic ulceration put into the hands of the neurologist for electrical treatment have been cured by the galvanic plates, or the use of the battery current. Electricity is very effectual in stimulating healthy granulations.

Faradization is used in the wards to improve the condition of palsied muscles; and central galvanization is employed chiefly in spinal affections.

METALLOSCOPY AND METALLOTHERAPY.

Numerous experiments in metalloscopy and metallotherapy have been made in the Nervous Wards, only a few of which can be alluded to at present.

In one case of brain tumor with partial anæsthesia of the left leg, a small zinc plate applied to this limb in an hour caused a sensation which was described by the patient as being like that produced by the "battery," referring to a faradic instrument. Other metals were tried, but had no effect. The salts of zinc were used without success, iodide of potassium being the only remedy that seemed to help the case.

Some curious results were obtained in a number of cases of marked anæsthesia from hysteria and spinal disease, to two of which reference will here be made.

One case was that of an unmarried woman, aged 29, supposed to be an example of hysterical paraplegia and anæsthesia. On two occasions plates of zinc, iron, copper, tin, silver and gold, of about the same size and weight, were placed on different parts of the body simultaneously; at other times the applications were varied—sometimes one plate, sometimes two or three were used. Many trials were made, the patient being blindfolded, and different locations being selected for the same plate. In five instances the patient picked out the zinc plate in from twenty to forty minutes, saying that she felt under it a sensation which she described as tingling, or as like "pins and needles." Twice she referred similar, but weaker, sensations to the plate of iron, but other metals gave no result.

Sensation was temporarily improved, muscular power was apparently increased; and the anæsthetic limbs bled more freely, on pricking them with needles, after the zinc was applied, until the peculiar sensations described were called forth. This patient was kept upon the use of valerianate of zinc for six weeks—sensation, motion, and her mental condition improving. Subsequently, however, she relapsed.

A second case was that of a man, aged 28; an advanced case of sclerosis of the posterior columns, with almost absolute anæsthesia of the lower extremities. After carefully testing the condition of sensibility and of the circulation, a small zinc plate was applied to the right calf, and a silver plate of the same size to a corresponding part of the left leg. In thirty minutes he began to have a sensation as if needles were pricking him under the silver plate. Two or three minutes later he had a similar, but weaker, sensation under the zinc on the right limb. The plates were kept on ten minutes, during which time he had four alternations of sensation in the two sides. When the pricking sensation was present under the silver plate it would be absent under the zinc, and *vicè versa*; but it was in each instance much more decided under the silver. On removal of the plates electro-sensibility was decidedly improved. No change of sensibility to the æsthesiometer or state of the circulation was produced. The symptoms in this case were decidedly ameliorated by both nitrate and oxide of silver, but were not permanently benefited by any treatment.

Dr. Mills does not believe that the theory of "expectant attention" will explain satisfactorily all the phenomena which result from metallic applications. Patients do certainly sometimes exhibit metallic idiosyncrasies—whatever may be the explanation. Anæsthesia, even when the result of organic disease, can be temporarily removed by applying pieces of metal. He has observed that two metals will sometimes produce similar effects on the same individual · but, even in these cases, he has always found that one of the two will give rise to more decided sensations, and will be more positively effectual in removing the anæsthesia.

In regard to internal metallotherapy, it is somewhat difficult to arrive at a decision. Irrespective of metalloscopic investigations, the value, in chronic spinal diseases, of the preparations of zinc, silver, and other metals, has long been known. They can also be used with advantage in cases in which no effect is produced by external applications of metals. The salts of silver and zinc will undoubtedly bring about amelioration of serious symptoms in cases in which these metals, when applied to anæsthetic limbs, will be selected by patients in preference to others, because of the peculiar sensations which they cause.

MASSAGE AND SWEDISH MOVEMENTS.

Both massage and Swedish movements are employed to a considerable extent, some of the nurses being trained for this work. Massage is found to be of benefit, even in old cases of paralysis, serving to keep up nutrition and temperature, and preventing trophic changes. In neuralgic and hysterical cases it also often proves of great service.

In the same room in which the permanent electrical instruments are kept, are some simple forms of apparatus for the movement treatment, such as a crossbar adjustable at various heights, a leaning cylinder for exercising the muscles of the trunk, a stool of the proper height and size for sitting movements, and a lounge or couch so hinged as to be capable of being inclined at various angles. The patients are taught to practice movements with or without assistance, according to the nature of the case.

A movement treatment, without apparatus, is also often

used. The kinds of movements usually resorted to, without
appliances, are the passive, or the duplicated active. Systematic passive movements are employed for the purpose of
preventing, as far as possible, atrophy and deformities. Joints
are kept in a healthier condition through the agency both of
massage and these passive movements. Duplicated active
movements are used in those cases in which the loss of power
in sclerotic or paralytic patients, for instance, is not absolute.
In conjunction with faradization this method of treatment
often results in the marked improvement of the paralyzed
limbs, palliating symptoms, and improving circulation and
nutrition even of palsied limbs.

THE ACTUAL CAUTERY.

The actual cautery, either alone or conjoined with other
remedies, is frequently resorted to in the treatment of epilepsy,
and of chronic spinal diseases. The ordinary cautery-iron,
with a button shaped like the blunt end of an olive, has
usually been employed, but recently the hospital has obtained
a Pacquelin cautery, in which the vapor of pure benzine is
forced by an air-blast upon a piece of hot platinum. Superficial applications to the nape of the neck, or along the spinal
column, are made every two or three days. The intervals
between epileptic seizure have been extended from days to
months by the use of the cautery

THE TREATMENT OF SYPHILITIC BRAIN DISEASE.

The wards are nearly always well supplied with syphilitic
affections of the brain and cord. Iodide of potassium in
energetic doses is largely employed. Mercurial inunction has
also been extensively tested, and in a few instances with
striking results. From ½ a drachm to 1½ drachms of mercurial
ointment is used daily, or every other day, the treatment being
persisted in until some effect is produced, or good reasons

arise for its discontinuance. Before inunction, the parts to which the ointment is to be applied are well sponged with warm water. Strict attention is paid, at the same time, to diet and hygiene.

TREATMENT OF SPINAL SCLEROSES.

For the various forms of spinal scleroses, and particularly for posterior spinal scleroses, or locomotor ataxia, the salts of silver—the nitrate, phosphate, or oxide—are generally found to be the most efficacious internal remedies. They are used in doses of from $\frac{1}{8}$ to $\frac{1}{4}$ a grain, and are often combined with some bitter tonic, as the extract of gentian or quassia. Electricity, in the form of moderately strong galvanic currents, is also much used; stabile currents to the spine, and labile currents to the limbs being the most common methods of application. Early in posterior scleroses large doses of ergot are often prescribed.

THE TREATMENT OF CEREBRAL AND SPINAL EXHAUSTION.

Preparations of phosphorus are used in the treatment of cases which show signs of cerebral or spinal exhaustion. A favorite preparation of this substance is the oil of phosphorus of the Prussian Pharmacopœia. This oil is administered according to the following formula, which is also used at the Hospital of the University of Pennsylvania:

℞ Olei phosphorati... ℳxvj.
    Olei gaultheriæ.. ℳviij.
    Mucilag. acaciæ..q. s. ad f℥j.   M.
    Sig.—One to two teaspoonfuls three times daily.

The oil of phosphorus itself can be prepared by the following process: "Into 5 fluid drachms of pure almond or olive oil, contained in a glass flask, drop 3 grains of transparent phosphorus. Place the whole in a water-bath at 175° F., and agitate until dissolved."

CALABAR BEAN IN DEMENTIA PARALYTICA

Calabar bean is prescribed in dementia paralytica, cases of which, in the early stages of the disease, sometimes find their way into the Nervous Wards. If not promptly relieved, they are transferred to the Insane Department. Pills of the ext. physostig. venenas, each containing $1/6$ to the $1/3$ of a grain, are given three times daily, the treatment being persistently continued and the effects of the drug constantly watched. Rest, nourishment, and counter-irritation to the head or nape of the neck are conjoined with the calabar bean

GENERAL NOTES.

Cannabis indica, hyoscyamus, conium, morphia, chloral, and bromide of potassium are used to fulfil various indications, such as tremor, headache, sleeplessness, mental symptoms, etc

APOPLEXY.

In the treatment of the apoplectic state the patients do not stand depletion well. Bleeding is seldom employed. Supporting measures are often found to be necessary to carry the cases successfully through the attacks.

THREE INTERESTING CASES OF SPINAL DISEASE.

CASE I.—The patient, a sailor, was brought into the hospital with a history of a fall of thirty feet from the rigging of his ship while at sea. He fell partly on his head and partly on his back. When picked up he was unconscious, and remained so for fifteen minutes. Upon regaining consciousness he found that his right arm and leg were entirely paralyzed, and the left arm almost entirely so. The patient received no treatment whatever while at sea.

Since admission the patient had been carefully examined

and found to be free from any disease of his heart, lungs, liver, or kidneys. The ophthalmoscope revealed a slight optic neuritis, but not enough to indicate any serious disease of the brain. When he walked he carried his head somewhat forward. The effort to straighten the head gave the patient pain. At the point of junction of the cervical and dorsal spine some thickening and induration were found. The right arm had not entirely recovered its power. He was unable either to extend or close the fingers of that hand. There was some atrophy of the muscles of that hand. The circulation of the same limb was also found to be very defective. The left hand and arm exhibited the same conditions, but to a lesser degree than the right arm. The man's walk was peculiar. The right leg was stiff and trembled when he walked. Upon stripping the limb the feet were found to be abnormally extended. Attempts to bring the foot up to a right angle were attended with great trembling of the member and marked tension of the tendo Achillis. The tendon reflex of the patella of both legs was most marked. There was no decided impairment of the sensation of either touch or pain in the feet. Contraction of the muscles under the faradic current was slightly impaired in both legs and arms. Now and then trembling could be produced by pressure upon the lower part of the spinal column.

In debating the case, two difficulties were encountered. First, in connection with the nature of the original injury; and second, as regarded the present nature of the disease. Concerning the first point, concussion of the spine seemed hardly possible, since it could not account for the paralysis of the muscles of three extremities. It seemed more probably to have been a case of apoplexy of the cord. It was concluded, however, that the effusion of blood could not have been a large one, since there was no paralysis of the left leg, no urinary difficulty, and no pronounced tendency to the formation of bed-sores.

Concerning the second point to be decided, it was thought

by Dr. James H. Hutchinson that the symptoms then present pointed conclusively to an involvement of the lateral columns. This opinion was strengthened by the presence of spastic muscular contractions. The absence of anæsthesia proved that the posterior columns could not be seriously affected, and the absence of analgesia seemed to show the same to be true of the gray matter. It was thought that simple concussion of the spine was sufficient to produce inflammation of the lateral columns.

As soon as the patient entered the wards a blister was applied to the seat of induration in the back, and bromide of potassium administered internally. When it was seen that inflammation of the cord existed, $1/16$ of a grain of bichloride of mercury was ordered four times a day. Later he was put upon gr. x of potassium iodide thrice daily, which dose was subsequently doubled. It was remarked, in connection with the early history of the case, that ergot and belladonna would have been the proper remedies to employ immediately upon the reception of the original injury.

For the future, occasional blisters were ordered and rest was enjoined. Strychnia and electricity were regarded as injurious.

CASE II.—A farmer, with a good family history, who had spent a day in very hard work, and during the following night had been much exposed to cold and wet. This imprudence, which took place eleven months before his admission to the hospital, was followed by loss of power in the lower extremities and complete paralysis of the bladder.

For two or three months after this period the patient had shooting pains in his legs, and was much troubled with nocturnal delirium.

The man was carefully examined by Dr. DaCosta after his admission, and found to be suffering from marked paraplegia without involvement of the rectum or bladder. The strong faradic current elicited no response whatever in either leg, except from the flexors of one of the big toes; using the continued current the same results were observed.

The case was regarded as a typical one of acute spinal paralysis—a rare form of disease in the adult—in which the lesion was in the anterior horns of the spinal cord. The location of the lesion was thought to account for the entire loss of electro-muscular contractility, while sensation was so little impaired. It was also thought to account for the absence of ectal and vesical difficulties and of bed-sores.

CASE III.—A shoemaker, who stated that his troubles dated back to an attack of rheumatism in his legs. This rheumatic attack lasted about three weeks. Pains in the back and legs were associated with rheumatism.

When admitted to the hospital the man complained greatly of pains in the limbs and back, and of loss of power in the lower extremities. He still walked with difficulty and pain, but the local tenderness and discoloration had gone. It was difficult to decide whether the pains complained of were due to the rheumatism in the extremities. It was concluded that they were not. The man improved very rapidly under the use of the iodide of potassium and ergot, and was discharged entirely cured, as it was supposed. Ten days afterward, however, he was again admitted. Upon questioning him closely it was discovered that he had spent all the time since his discharge in walking about the city in search of employment.

After his return to the wards he began to lose power in his legs steadily. The pains in his back and legs returned, and he spoke of a feeling of great constriction around his waist. Still later the legs began to atrophy, and still the loss of power remained. The electro-muscular contractility was very much diminished, while the sensibility seemed to be slightly deficient in both of the lower limbs. The capillary circulation was very defective, the blood circulating very irregularly in the superficial tissues. Reflex sensibility still remained intact. The muscular sensibility had gradually increased.

It was concluded very early in the progress of this case that it was one of rheumatic paralysis of the cord, or rheumatic spinal myelitis.

Anatomically speaking, it was pointed out that the lesions in Cases II and III were the same, both being located in the anterior horns, but in Case II there was no rheumatism, and the spinal paralysis came on at once; while in Case III the rheumatic origin was plain, and the spinal complications came on at a later period. The treatment of Case III was by gr. xv of the iodide of potassium thrice daily, with a little iron and locally friction.

In the discussion of Cases II and III some very interesting points were brought out. Particular attention was directed to the rapid atrophy of the muscles of the lower extremities in both of the cases. This atrophy was emphasized as being the most typical and constant symptom of this class of affections, the rapidity with which this atrophy progresses being in proportion to the acuteness and persistency of the attack.

Another fact, to which particular attention was directed, was that this atrophy, as a general rule, is not permanent. In cases where recovery has been more or less complete, the limbs are found to have regained, to a greater or less degree, their normal shape and power, the completeness of the return of the limb to its normal shape being dependent, of course, upon the completeness of the patient's convalescence. Complete restoration to power and health was regarded as rare. It was argued that the real extent of the damage done depended upon the number of trophic cells which had been destroyed. If enough of them remained intact, when the morbid process had ceased, to minister sufficiently to the supply and nourishment of the atrophied muscles, it was easy to restore their for a time lost functions by means of proper nerve-food, friction and electricity in the shape of galvanism, and in the later stages by the hypodermic use of strychnia.

The question arising as to whether the tape-measure was the only means of judging definitely of the condition of the affected muscles, it was pointed out that when it was found upon trial that the faradic current when applied to the affected muscles began to give better results, and when the muscles began to respond and contract, though feebly, it might be accepted as a sure sign of returning health, for it proved conclusively that the muscles were becoming more active, and that the paralysis had reached its height.

It was thought that the same conclusion could not be drawn from the use of the continuous current, the muscles responding to this current so soon that it could not be regarded as any gauge at all.

Attention was called to the facts of the entire absence of rectal and vesical paralysis, and of bed-sores, in Cases II and III.

In one point it was shown that these two cases had not been typical ones of their kind, viz., in so far as the reflex nervous functions had not been in the least impaired in either case.

The treatment was limned out by Dr. DaCosta, as having consisted at first of local blood-letting in the neighborhood of the spine, occasional purging, and large and continuous doses of ergot. This at first; later the indications were met by large doses of the iodide of potassium, and by the application of systematic friction to the legs. Later still, it was thought that small doses of strychnia should be administered hypodermically, and when muscular motion returned, that the faradic current should be employed.

CHOREA.

Dr. Wharton Sinkler has controlled the twitchings of patients with this disease by the continued hypodermic injection of from 3 to 5 drops of Fowler's solution. In some of the instances, the injections were followed by a good deal of local irritation.

Dr. H. C. Wood has treated successfully several cases of this disease, with a saturated tincture of the rhizome of dracontium. The dose at first is 60 drops thrice daily, and it is gradually increased to 90 drops. Particular care is taken that the preparation of the drug used is a reliable one.

PARALYSES OF THE RECURRENT LARYNGEAL NERVE.

Nerve tonics, such as strychnia, phosphorus, iron, and cod-liver oil, are used. In the so-called instances of hysterical and nervous aphonia, a sponge-probang is moistened and brought into contact with the vocal cords, which are thus thrown into a state of spasm and so brought together. The same effect is produced in other cases by sprays of ether projected upon the part, or by inhalations of pungent substances.

When all else fails, electricity is employed and is applied directly to the paralyzed muscle.

In using electricity, the plan pursued by Dr. Cohen is to place one electrode directly over the crico-thyroid ligaments outside, and then to carry the other electrode into the larynx and place its point in contact with the cords, or between them. The electrode in the throat is only kept in position a few seconds at a time. Cures are often effected in this way by a single application.

Where intra-laryngeal electric excitation is not possible, the percutaneous method is tried. This consists in passing a current from one side of the neck to the other, and so through the larynx. Or, the patient is placed upon the insulating stool and a spark is drawn from the cricoid cartilage with the knuckle.

In cases of hysterical aphonia a cure is often effected by the mere introduction of the laryngeal mirror, the patient being given to understand that this is the curative procedure.

Another method often tried is to stand behind the patient and grasp the thyroid cartilage between the thumb and forefinger, while at the same time the middle finger is placed under the cricoid cartilage, pulling it up and in front of the thyroid. In this way the vocal cords are stretched and made tense, and so caused to vibrate by means of the inspiratory current.

TETANUS.

Systematic feeding of patients with liquid and strengthening food at short intervals, has been employed by Dr. H. C. Wood with very good results. The food is given at intervals of every two or three hours, and consists mainly of milk, with a small quantity of alcohol. In severe cases all solid food is avoided. As regards medicines, the patient is brought well under the influence of the bromide of potassium by an initial dose of from 2 drachms to ½ an ounce, to be followed by ½ a drachm to a drachm every three or four hours. To force sleep at night, 30 grains of chloral are given at bed-time with some opium. Chloral is also used, when necessary, in the day-time. Nitrite of amyl and chloroform are never used steadily, but are employed from time to time to stop violent spasms. Where there is much cerebral congestion, a blister is applied to the nape of the neck.

WEAVER'S, VIOLINIST S, AND IRON PULLER'S PALSIES.

CASE I.—W. B. C., by profession a *violinist*, has been a steady player for twenty years, averaging six hours out of every twenty-four. Patient presented himself for treatment January 27th, 1875, saying that one year before that date he first noticed a spasm of the flexors of the ring and little fingers during an attempt to lift them from the cords of the violin. When playing he had pain in the ulnar region of the arm. Used to play first violin; but for a year had been obliged to play second violin.

Sensation and electro-muscular contractility remained unimpaired.

He holds neck of violin between the thumb and forefinger, touching the cords with the other fingers. He occasionally experiences a momentary cramp in the fourth and fifth fingers in lifting them off the cords. *Was able to play the piano without any difficulty.*

September 27th, 1875 —Trouble still continued in the third finger of the left hand, which hung on his violin string when it should be raised off it. Doubtful of any improvement. Advised galvanization of flexors and extensors, one pole in axilla, the other stroked over the muscles.

CASE II.—W. V., æt. 35 years, a *heater* in a rolling mill, who worked in an iron foundry at Catasauqua, pulling the iron in and out of the furnace.

Patient had been married and had four healthy children. He denied syphilis, and had always enjoyed good health up to the date of his present troubles. He came for treatment on March 25th, 1872, with the statement that after working three successive terms at the furnace, and sleeping from seven to nine A. M., he went to a pic-nic. The next day when he woke up he found his hands swollen and stiff. This was on a Friday. He did not resume work until Monday, by which time the swelling of the hands was less marked. He worked steadily for a month, the strength in his hands failing, the grip becoming weak, with a feeling of numbness but no loss of sensation.

During this time his wife died, and he was much distressed. For a month after her death he worked, and at the end of that time was obliged to stop. Since then the strength in his hands has steadily failed, commencing in the hands and extending up the arms.

There was some pain in the small of the back, some difficulty in articulating, and also in "hawking." No dyspeptic symptoms. Bowels very irregular. Passed urine but once during the day, and it never drippled away. No blue lines on the gums.

There was absolute loss of power in the deltoids, though the fibres of the muscles contracted. No power in biceps, and the greatest palsy in the flexors of the fingers and hands. Able to raise wrist, and to pronate and supinate arms. Loss of power in the triceps muscles not so marked as in the others, the right triceps being stronger than any other arm muscles

Pectoral muscles weak. All the muscles were about equally atrophied, and to a considerable extent.

Sensibility and localization good. Could test the distance of points well.

Some loss of power in uvula apparent. Both right and left arm measured 7½ in. at a point three inches below the elbow.

Electric condition.—The secondary induced current one cell, three inches, moved both deltoids, most marked contraction being obtained by placing the positive pole over the brachial plexus, and the negative over the muscle. Electro-muscular contractility remained in all the muscles, but was diminished in degree except in the flexor communis digitorum. Electro-muscular sensibility existed in both arms.

℞. ext. ergot. fl., gtt. xx., and potass. iodid., gr. v., t. d.; also advised the use of galvanism.

April 2d.—The dose of the medicine was diminished one-half in consequence of its purging him. One week later it was found necessary to discontinue it entirely for the same reason.

April 30th.—After eighteen applications of galvanism there was no improvement, and, returning home, he died about the middle of May.

CASE III.—J. B., æt. 26 years, a weaver by occupation. His family history has been good; so also has his previous health been. Patient has been a weaver since he was fourteen years of age, using his left hand as much as the right in weaving. When fifteen years old he began to be troubled with spasmodic pronation of the left hand. He stopped work for a few weeks, and upon returning was not troubled with it again until he was twenty-one years of age, when it re-appeared, and has continued permanent ever since, except when he could stop weaving temporarily.

The spasm consists of a sudden pronation of the left hand, with a general tremor of the arm, caused by any voluntary exertion—a large, irregular tremor. Any movement of the hand with the arm flexed is more difficult to perform than with the arm extended. Patient is unable to comb his hair *backwards* with the left arm.

He was naturally left-handed, and when he first presented himself for treatment at the clinic, the dynamometer registered with the left hand 135, right 150.

Sensation was unimpaired in both hands. There was no pain in the back of the neck; but a "bruised" feeling in left arm after exertion. Advised change of occupation and galvanism three times weekly to arm.—*Dr. S. Weir Mitchell.*

FACIAL PARALYSIS, FOLLOWED BY SECONDARY SPASMS AND CONTRACTURES.

R. S., æt. 46, a married woman, came under observation late in the winter. Three weeks before, without known cause, except that she was at times considerably exposed to the changes of weather, she suddenly became paralyzed on the left side of the face. Her menses had been irregular for six months, and she had not been feeling very strong for a year.

Examination showed complete paralysis of the muscles supplied by the external third of the facial nerve. The left side of the face was smooth. She could not wrinkle the left half of the forehead, nor close the left eye. The nose was drawn very slightly to the right, and the mouth in the same direction and upwards. She could not pucker the mouth. Her speech was somewhat indistinct, and, when eating, food often lodged between the left cheek and the teeth, causing annoyance. The conjunctiva of the left eye was usually injected, and she complained frequently of pain and discomfort in this eye.

The velum palati and uvula were unaffected. She had no loss or disturbance of taste on either side of the tongue. Smell was slightly impaired. Hearing in the paralyzed side was normal, being neither abnormally acute, nor dull nor lost, and she had no noises in the ear.

The muscles responded promptly to both the faradic and galvanic currents, and the response to faradism, which was used in the treatment, continued good.

Iodide of potassium was prescribed in doses of 10 grains three times daily; and faradization with a current just sufficient to produce muscular contraction was employed every other day. Positive and progressive improvement took place, the patient gradually recovering power in all of the affected muscles. After six weeks of treatment, however, a slight twitching of the left angle of the mouth began to be noticed. In a few days, this angle of the mouth was drawn up almost constantly. The cheeks and lips began to be pressed against the teeth, and she complained of a band-like feeling around the paralyzed side of the mouth. She had, in short, an irregularly distributed, but marked condition of spasm in most of the muscles recovering from the paralysis. Now and then a succession of rapid twitches would be noticed in the muscles about the angle of the mouth. Less frequently similar twitchings were observed in the orbicularis palpebrarum muscles, especially in its lower fibres. Any effort, under the influence of the will, to use certain muscles or groups of muscles of the left side, would cause a curious appearance of distortion and grimacing, owing to the irregular spasmodic actions that would ensue. When the mouth was pulled upwards and outwards, for instance, the eyelids would be pressed together and the digastric muscle would be called into play. The patient was constantly annoyed by an unpleasant feeling of "drawing." After this condition had lasted from one to two weeks the left half of the orbicularis oris and the zygomatic muscles began to feel hard and hypertrophied.

On the appearance of the spasmodic symptoms, the use of the faradic current was transferred to the healthy or non-paralyzed side and a weak galvanic current, uninterrupted, was applied to the zygomatic, orbicular and other muscles, which were the seat of the tonic spasm or contractures. Bromide of potassium and valerianate of zinc were administered internally, and belladonna ointment was frequently rubbed into the face. Under this treatment the patient improved, and she was discharged much relieved, but not entirely cured, as *some* contracture still remained.

The secondary contractures and spasms observed in this case have been noted by Dr. Mills as occurring in varying degree in a number of cases of facial paralysis. They have been studied by Duchenne, Remak, Hitzig, Erb, and others. No very satisfactory explanation of their occurrence has been offered. Hitzig refers the symptoms to an abnormal excitability of the medulla oblongata, which becomes developed in a still unknown manner, in consequence of peripheral facial paralysis. (Erb in Ziemssen's Cyclopædia, vol. XI, p. 109.) Dr. Mills believes with Erb, that the condition is not one of "electrical muscle tetanus," induced by electric treatment, as it occurs in cases in which no electric treatment has been employed. Transient states of spasms in cases of this kind are by no means uncommon. Assuming the doctrine of the localization of motor centres in the cerebral convolutions to be true, it may be that the special cervical centres for various facial movements, temporarily quiescent during the existence of the paralysis, as recovery takes place begin to act in an irregular and abnormal manner.

Whatever may be the true theory of the production of these secondary spasms and contractures, experience has proved that they are best treated by the conjoint use of internal and external measures. The bromides and preparations of zinc, cimicifuga, hyoscyamus, arsenic and similar articles, should be given with a view of quieting and controlling the nervous centres, while at the same time, the patient's strength is sustained by mild tonics and abundant nourishment. A weak, stable galvanic current should be used for its antispastic effect directly to the nerves and muscles. Mild faradization of the antagonistic muscles of the healthy side can also be resorted to, but it should be used with caution and not too often. Detmold's mechanical treatment of facial paralysis can be applied with advantage to the unaffected side with the view of preventing and controlling contractures upon the opposite side. This is carried out by taking a piece of tin wire, or some similar material, and bending it at both extremities, so that one end can be passed over the ear and the other hooked into the angle of the mouth, thus affording a support and preventing the drawing to the other side. Hypodermic injections of sulphate of atropia, gr. $1/100$ to $1/60$, using fine needles, can be thrown into the spastic muscles; and here, as in so many other cases of spasm, superficial applications of the white-hot iron to the back of the neck, or over the muscles, may prove of signal service. Cases of this kind sometimes obstinately resist all therapeutic measures.

FACIAL PARALYSIS, WITH LOSS OF HEARING, TINNITUS, AND PECULIAR VERTIGO.

B. G., æt. 52, a widow, had been healthy until twelve years before admission, when she had a severe attack of rheumatism, chiefly involving the knee joints. Four weeks later she began to have violent pain in the right side of her head, which subsequently extended to the top, and then to the left side. She suffered from these head pains, with but slight intermissions, for two years, when total paralysis of the left side of the face made its appearance, and was accompanied by loss of hearing and by noises in the ears. Spells of giddiness of a peculiar character came on with the paralysis. On attempting to walk she would soon be compelled to run, and on getting into a run, she would fall on her face, unless she could stop herself by taking hold of some object. She felt as though she, herself, and the floor under her were going around. These vertiginous spells gradually grew less frequent and severe as the patient's general health became better. She has never had a discharge from the ear, but has suffered from more or less headache ever since the facial palsy occurred. For three months she had double vision. For six months she had some difficulty in swallowing, fluids sometimes being regurgitated through her nose. On admission the right side of her face and both arms and legs were found to be unaffected by paralysis, but the left side of the face showed marked paralysis and atrophy. The usual lines and furrows were obliterated, and the cheek sagged downwards. No movements could be performed by any of the muscles supplied by the left facial nerve. Lagophthalmos was marked, the left eyelids remaining wide open. The lower lid was slightly ectropic. The conjunctiva was generally somewhat injected. She had full control of the movements of the left eye, which she could move in any direction. A peculiar motility of this eye was noticeable. It was almost constantly jerking or wabbling upwards and downwards, going slightly outwards in its ascent. Sight in this eye, when it was fixed, was good. Her nose and mouth were drawn far to the right. She had no control whateve over the paralyzed muscles; she could not dilate the nostril, raise the lip, draw upwards or outwards the angle of the mouth, etc. She could speak distinctly and protrude the tongue without visible deflection. The uvula pointed slightly towards the right, and the velum hung lower on the left than the right side. On touching it with a probe it retracted upwards and towards the right. Taste was generally defective, but it could not be made out distinctly to be abolished on the anterior part of the left half of the tongue. She complained of dryness of the mouth. Hearing in the left ear was entirely gone, and she had constant noises in this ear. Smell was defective, but no differences could be made out between the paralyzed and healthy side. Sensation, as determined by the æsthesiometer and the faradic battery, was slightly, but undoubtedly, diminished on the affected side. Both farado-contractility and galvano-contractility were also much diminished. Reflex movements could not be produced by irritating the skin of the face. The surface temperature was carefully taken in the middle of each cheek; it was 91.8° F. on the left or paralyzed side, and 95.8° F. on the right. At times she still had spells of giddiness. On going up stairs she became giddy, but was never troubled in this way on coming down. Her appetite was poor; her bowels were constipated; she was frequently troubled with nausea, particularly in the mornings. The nausea was not accompanied by vertigo. She was nervous and irritable, and her general condition as regards strength was not good.

The symptoms presented by this case were interesting, and some of them unusual. They indicated an extensive lesion, probably a rheumatic or syphilitic exudation or tumor compressing the facial nerve at the base of the skull. Destructive disease of the petrous portion of the temporal bone might also explain the case. The peculiar form of vertigo is worthy of note. The facial nerve, in the first part of its course, passes forward, resting on the cerebellar crus, and it might be considered whether an involvement by the disease of this arm of the cerebellum, or of the cerebellum itself, had not some agency in the production of the vertigo, and the tendency to run and plunge forward. Auditory vertigo does not usually take the form presented in this patient.

The head symptoms—pain and vertigo—were greatly benefited by the use of iodide of potassium and quinine. Strychnia and carbonate of iron were also administered with the effect of improving the general condition of the patient. Massage and faradization and galvanization, both with the continuous and the interrupted current, were persistently employed directly to the nerve-branches and muscles, but with little benefit, as both nerve and muscle degeneration had gone so far as to preclude much hope.—*Dr. C. K. Mills.*

SPASM OF THE SPINAL ACCESSORY NERVE.

Dr. Wood's treatment is by means of hot irons applied to the nape of the neck and immediately over the contracted muscles. The heat used is intense, and is only very lightly applied, so as not to produce a very deep eschar. Together with this cauterization the iodide of potassium is administered.

SEXUAL EXHAUSTION.

When this condition is a result of masturbation, as a first and most important step the patient is persuaded to stop the practice at once and forever. Then the hygienic treatment is in order. Enough exercise is taken each day to produce decided fatigue. Meat is forbidden as a general article of diet, and the patient lives largely upon farinaceous food. A general meat diet has been proved to throw a great strain on the kidneys.

All kinds of exercise which irritate the genital organs are strictly forbidden by Dr. H. C. Wood. As sleeping on the back provokes emissions, the patient is advised to always sleep on his side on a hard bed, with as few covers as the weather will allow. All sexual literature and theatrical scenes are avoided.

As regards medical treatment, bromide of potassium is given in doses of from 20 grains to half a drachm thrice daily, or in the place of the bromide 5 grains of brominated camphor are given three times a day in emulsion. Along with this soothing treatment iron and some bitter tonic are with propriety employed. In some instances ergot, by relieving the congestion of the spinal centres, does great good.

The treatment of sexual exhaustion from excessive venery is about the same as that just sketched out. A specific remedy is phosphorus. Ergot is also always used, where there is numbness or prickling of the limbs.

In impotence, with spermatorrhœa, the following prescription has been found to act like a charm:

R   Tinc. canthar .... gtt. vj.
    Tinc. ferri chlor .... gtt. xv-xx.   M.
Sig.—Thrice daily in water.

PARAPLEGIA DUE TO A SPINAL SPRAIN.

The most important indication is thought by Dr. Levis to consist in producing a powerful counter-irritant effect along the spine, with a view of removing, if possible, the plastic effusion upon which the paraplegia depends. With this indication in view, the most powerful counter-irritant known, which is the iron at a white heat, is employed. In using the hot iron upon a patient, he is placed under the influence of an anæsthetic and six issues are made, three upon either side of the spine, in the lumbo-dorsal region.

There is a nice little point about the after-treatment of the cauterized spots, which has not, as a rule, received the attention it merits. If, after the hot iron has been applied, the parts be left exposed to the air, or even the ordinary cold water dressing be applied, the patient will suffer a great deal of pain. But if pure carbolic acid—which has simply been deliquesced by the application of a moderate heat, so that it may be conveniently spread with a camel's-hair pencil—be applied to the cauterized spots, it produces such a marked anæsthetic effect as to take away almost all the pain which would otherwise harass and distress the patient. This is a little point, but one which, if observed, relieves considerable suffering. The parts are protected by the ordinary patent lint and carbolized oil dressing, which is covered with waxed paper, and secured in place by adhesive strips.

During the period of inflammation, in spinal lesions of this character, the bromide of potassium is indicated; but later on in the affection, strychnia is the agent upon which most reliance is placed.

MISCELLANEOUS.

THE INTRAVENOUS INJECTION OF MILK IN FUNCTIONAL AND ORGANIC ANÆMIAS.

A female, æt. 33, was admitted to the University Hospital in the spring of 1878, suffering from extreme anæmia and spinal irritability, which had been induced by a series of depressing causes, miscarriages, hemorrhages, mental strains and nervous shocks, the latter requiring a large amount of morphia in their management. Other forms of treatment having been assiduously persevered in without any good effect, it was determined to inject milk into her veins. On June 20th, Dr. Charles T. Hunter opened her right median basilic vein and injected five fluid ounces of fresh cow's milk, the cow being milked in an apartment adjacent to that in which the patient lay. The immediate results of the operation were most striking. Almost as soon as the milk entered her veins the patient's face assumed a deep purplish color, her conjunctivæ became injected, and she clutched at her throat. These symptoms disappeared, however, upon lowering the funnel and allowing the milk to flow in more slowly. In a few minutes a very marked capillary injection appeared in the palms of her hands, her face and the back of her neck. At the same time two attacks of uticaria came on at intervals of eight minutes, and were gone in five minutes after the first appearance of the rash. Half an hour after the operation a chill supervened, lasting fifteen minutes. Before the operation the pulse was 108. After the milk was injected it ran up to 150, being 128 when the uticaria appeared. Two hours and a half after the injection the temperature ran suddenly up to 103° F., and then fell again. The patient passed a very comfortable night.

June 25th.—No marked improvement.

June 27th.—The patient was again operated upon, the same quantity of milk being injected. Previous to the operation 20 grains of quinia were given. [This same amount of quinia had been administered previous to the first operation.] No sooner had a fluid drachm of milk entered the circulation than great capillary congestion followed, and the patient complained of bursting pain in her head. A severe cramp seized her, and in the course of fifteen minutes her menses appeared. Again two attacks of uticaria supervened.

June 29th.—The patient was much stronger.

July 17th.—Milk to the amount of 6 fluid ounces was again injected. The injection was followed by the same symptoms. From that date the patient's condition steadily improved. She gained in color, and her nervous phenomena disappeared entirely.

THE ASSIMILATION OF COD-LIVER OIL.

Some months ago Dr. Wm. H. Bennett instituted a series of experiments regarding the assimilation of cod-liver oil by the system. The stools of a number of patients, placed upon the daily use of this article, were carefully examined, and it was found that in the majority of cases these stools were oily, showing that the cod-liver oil had simply passed through the alimentary canal without absorption. Where the stools were oily the patients appeared to have derived no benefit from the use of the oil; but in the few cases where the stools were not oily, and where the oil had consequently been absorbed, the patients had grown fat.

Quite recently Dr. Milton Miller, the medical resident, at the Episcopal Hospital, has conducted a second series of experiments of the same nature, and has obtained like results.

SLEEPLESSNESS IN UTERINE DISORDERS.

When patients complain of nervousness or of sleeplessness, the bromide of potassium is given either alone or in combination. A cheap and efficient mixture is the following:

℞ Pulv. ferri sulph. exsic..................................gr. xxx.
    Potassii bromid.
    Rad. columbæ contûs........................āā ℨj.
    Aquæ bullientis..Oj.
Steep for 24 hours and then strain.
SIG.—One tablespoonful in a wineglassful of water just before or just after each meal.

In other cases the following is used:

℞ Elixir. humuli..f℥j.
    Elixir. ammon. valerian.
    Syrupi lactucarii..āā f℥ss. M.
SIG.—One dessertspoonful at bedtime, or during the day when needful.
—*Dr. Goodell.*

ATROPIA AS A PREVENTIVE OF PYÆMIC CHILLS.

In several instances of abscess of the liver and of pyæmia, Dr. Starr administered the 1/90 of a grain of atropia by hypodermic injection, and the 1/60 of a grain internally, to prevent the distressing chills consequent upon these conditions. This remedy acted like a charm. The effects of the dose or hypodermic injection given in the morning lasted through the following twenty-four hours. The same was the case with the belladonna-bath (tr. belladon., f℥ij; spts. frumenti, f℥ij; aquæ, f℥j. SIG.—To be applied to the whole surface of the body by sponge at bedtime.)

DIGITALIS BY HYPODERMIC INJECTION.

Dr. Starr has recently been employing digitalis hypodermically with much success in cases of advanced phthisis and of heart failure. The injections at first contained gtt. v of the tincture of digitalis and ℳx of water. The effects being negative, the amount of the digitalis was increased to gtt. x, with the most decided effects. The pulse fell at once from 120 to 105 in the minute. At one time as much as gtt. xv of the digitalis were injected with great advantage.

OPHTHALMIA NEONATORUM.

A solution of nitrate of silver (¾ of a grain to the ounce) is injected under the lids twice a day.

For the lids themselves the following is usually employed:

℞ Sodæ boratis.. .... gr. xij.
    Zinci sulphatis............. .... gr. j
    Aquæ camphoræ.. fʒj.
    Aquæ destillatæ.. f℥j. M.
Sig.—To be applied to the lids two or three times a day.

A plan of treatment highly recommended by Mr. Dickson, of London, viz., the injection between the lids every half-hour of a solution of alum (from 5 to 8 grains to the ounce), the strength of the solution to be gradually diminished as the case gets better, has also been tried with very gratifying results by Dr. J. F. Meigs.

INDEX.

A.

Abortion 97-100
Abscesses, The Hyperdistension of........ 71
Acne Rosacea........................37-38
Addison's Disease..................41-42
Amenorrhœa 104-105
Anæmias, The Intravenous Injection of Milk in Functional and Organic...................134-135
Anæmia and Chlorosis........... 90-92
Aneurism............................ 54-55
Angina, Acute........................ 47
" , Pseudo-Membranous..19-21
Aorta, Constriction of.............43-45
Aortic Valve, Patulous, Accompanying Chronic Bright's Disease....................................32-33
Apoplexy 116
Arthritis, Rheumatoid...........33-37
Atropia as a Preventive of Pyæmic Chills 136

B.

Brain, Unique Case of Injury to the.................................... 78-81
Brain, Compression of the......81-82
Brain Disease, Treatment of Syphilitic.....................114-115
Breasts, Gathered..............100-101
Bright's Disease, Patulous Aortic Valve Accompanying Chronic............................32-33
Burn followed by Contraction of Tissue..........................72-73

C.

Calabar Bean in Dementia Paralytica.................................. 116
Cancer, Epithelial.................... 82
Catarrh, Post-Nasal.............51-52
Cautery, The Actual................ 114
Cellulitis, Pelvic Peritonitis and, 85-86
Cerebral Exhaustion, Treatment of....................................... 115
Cervix Uteri, Cancer of........... 106
" " , Conical 107
Chills, Pyæmic, Atropia as a Preventive of...................... 136
Chlorosis, Anæmia and..........90-92
Chorea.................................. 121
Clavicle, Fractured..........65 and 81
Cod Liver Oil, The Assimilation of....................................... 135
Collapse................................. 47
Constipation, Habitual with Fissure of Rectum.................... 104
Cystitis in the Female........106-107
" " Male...............66-68

D.

Dementia Paralytica, Calabar Bean in................................ 116
Diabetes Insipidus................... 51
" , Saccharine............. 40-41
Digitalis by Hypodermic Injection.............................136-137
Douglas' Pouch, Hæmatoma in, 87
Dropsy, Tapping in.................. 70
Dysentery, Acute and Chronic, 48-49

INDEX.

Dysmenorrhœa 95
Dyspepsia......................... 49-51

E.

Eczema, Chronic 33-37
Electricity....................... 110-111
Elephantiasis Arabum, Nerve Section for......................61-64
Endometrium, Fungous Granulations of........................... 108
Erysipelas........................ 83
Eustachian Tubes, Inflammation of...............................28-29
Exhaustion, Sexual................ 132

F.

Fever, Contagion in Scarlet.....30-31
" Pleuro-Pneumonia following Typhoid................ 12-15
Fever, Puerperal 94-95
" Typhoid, Some cases of, and their treatment 5-10
Fever, Typhoid, some interesting points in the Diagnosis and Prognosis of 3-5
Fever, Typhoid, the treatment of.............................10-12
Fever, Thermic.................... 46
Fistula, Vesico-Vaginal, Closure of Vulva for...................... 108
Fracture below the Knee, Extension in........................ 81
Fracture, Intracapsular........... 81
Frontal Bone, Rapid Recovery from Depressed Fracture of, 71-72
Funis, The Treatment of the...... 100

G.

General Notes..................... 116
Gleet............................. 64
Goitre 49
Gonorrhœa in the Male............ 73
Graves' Disease................ 27-28

H.

Hæmoptysis........................ 40
Heart Disease, The Treatment of Organic.......................17-19
Heart, Hypertrophy of...........43-45
Hemorrhage, Post-Partum......88-90
Hemorrhoids, External............ 70
Hernia, Temporarily Irreducible...............................77-78
Hip-joint Disease................. 61

I.

Influenza......................... 55

K.

Kidneys, Albuminoid Degeneration of........................19-21

L.

Lumbago........................... 52

M.

Massage and Swedish Movement........................113-114
Menopause, The 92-93
Metaloscopy and Metalotherapy, 111-113
Middle Ear, Inflammation of the...............................28-29
Milk, Intravenous Injection of, in Functional and Organic Anæmias.....................134-135

N.

Nævus Materna.................69-70
Nasal Passages, Inflammation of, 28-29
Nephritis, Acute and Chronic Tubal............................. 56
Nephritis, Chronic Interstitial... 56
Nipples, Sore.................101-102

O.

Ophthalmia Neonatorum............ 137
Opium Habit, The.................. 54
Ovarian Cyst, Diagnosis of 95-96

INDEX. 141

Ovaries, Prolapse of the............ 107
Ovariotomy............................. 101
Ozena 19-21

P.

Palsy, Iron Pullers'............123-126
" Violinists'...............123-126
" Weavers'.................123-126
Paracentesis Thoracis, Indications against......................15-16
Paralysis, Facial, Followed by Secondary Spasms, etc......126-129
Paralysis, with Loss of Hearing, etc.............................129-131
Paraphimosis........................ 84
Paraplegia due to a Spinal Sprain 132-133
Patella, Fractured.................. 60
Perimetritis 87-88
Perinæum, Laceration of the...96-97
Peritonitis and Cellulitis, Pelvic, 85-86
Peritonitis Idiopathic..........21-23
Perityphlitis...................42-43
Phthisis, Laryngeal.............52-54
" , Night Sweats in........ 29
Placenta Prævia................... 102
Pneumonia 16-17
Psoriasis, Chronic.................. 42
Pregnancy, The Diagnosis of, 93-94
" , The Vomiting of.... 87
Prostatorrhœa with Stricture...68-69

R.

Rectum, Fissure of the, with Habitual Constipation.......... 104
Recurrent Laryngeal Nerve, Paralysis of........................ 122
Rheumatism, Acute and Chronic, 47-48

S.

Sciatica 38-40
Shock, Surgical 65
Sore Throat, Syphilitic..........76-77

Spine, Curvature of the............ 65
Spinal Disease, Three Interesting cases of..................116-121
Spinal Accessory Nerve, Spasm of the............................ 131
Spinal Exhaustion, The Treatment of........................... 115
Spinal Sclerosis, The Treatment of................................. 115
Spinal Sprain, Paraplegia due to a............................132-133
Sprains 73
Stomach, Dilatation of the....... 37
" , Ulcer of the............ 51
Stricture........................74-76
" with Prostatorrhœa....68-69
Swedish Movements, Massage and..........................113-114
Syphilis, Diagnosis of Congenital............................102-104

T.

Tetanus 123
Tinea Favosa........................ 38
Toe-Nail, Ingrowing............82-83
Tonsillitis, Acute................... 46

U.

Ulcers 65
Urœmia 56
Urethra, Foreign Body in......57-59
" , Caruncle of 105-106
Uterine Disease, Sleeplessness in................................. 136
Uterus, Anteflexion of..........86-87

V.

Vaginismus......................... 109
Vaginitis 108
Vertigo, Gastric 41
Vomiting, Two Interesting Cases of 23-27
Vulva, Closure of the, for Vesico-Vaginal Fistula................ 108

NOTES

OF

HOSPITAL PRACTICE.

PART II.

NEW YORK HOSPITALS

EDITED BY

SAMUEL M. MILLER, M. D.

PHILADELPHIA, PA.
SAMUEL M. MILLER, M. D., Publisher.

1880.

1

GENERAL DISEASES.

ACUTE PNEUMONIA.

Dr. Alonzo Clark has not lost a fondness for the lancet in pneumonia. He is accustomed to use cups after scarification, taking three, four, or five ounces from the affected side. He does not place any confidence in the calomel treatment, or in Laennec's treatment by tartar emetic, but is satisfied, when the temperature rises to 105° or 106°, that the safest and most efficient medicine is quinia, given in doses of 10 grains thrice daily. The application of cold to the surface of the body is thought to be disastrous. Dr. Clark does not allow any ice to be applied to any patient of his who has pneumonia. He, however, regards sponging with cold water as admissible, and is in favor of the employment of a bath at a temperature 20° below that of the body. His objection to veratrum viride is that, when it is given with any freedom, it is very apt to lead to collapse; aconite, he thinks, safer. He regards the alcoholics as utterly useless in the treatment of this affection.

Ten years ago, at Bellevue Hospital, pneumonia was quite uniformly treated with carbonate of ammonium, internally, and an oil-silk jacket, externally. The carbonate was given in doses of gr. v every three hours, or sometimes gr. x thrice daily. The muriate was occasionally substituted. Gradually, quinia came to be combined with the ammonium, while to-day, quinia has entirely taken its place in many wards. The quinia is given in doses of gr. x thrice daily, increased or diminished according to the fever. The oil-cloth jacket is still continued, and, if there is much pain in the side, a coat of iodine is ordered. This, with an absolutely recumbent posture, is all that is enjoined in many cases. Aconite has been used in five cases, of which one died. It is given in doses of ♏j every hour, until some effect of the drug, either in the relief of dyspnœa or fever, or the production of sweating, is brought about.

In 44 cases, quinine was used with good results. The method of administration is varied. Besides the routine mentioned, it is given gr. j every hour, or gr. v every four hours, or often gr. xl or gr. l, in one dose, then discontinuing it for a day or more. The antipyretic effect of quinine has not been sufficiently marked to make it clear which is the best way. Cold sponging has been employed with quinine. Of seven cases so treated, three died. The gradually-cooled bath has been used, and was at once given up.

Many cases that have been admitted into the hospital have had a record like the following: The patient is a tolerably strong man, in the third day of the fever. He has a temperature of 104°, respiration 50. His pulse is very good, and he feels pretty comfortable. He is given milk and eggs, and gr. xv of quinine. This is in the morning. In the afternoon he is weaker, his face is a little blue, he breathes faster. On listening to the lungs, moist râles, fine and coarse, are heard. He is beginning to have œdema. He is at once dry-cupped for fifteen or twenty minutes, during which time 150 cups are put on. The œdema has now disappeared. He is ordered ♏x of tr. digitalis every three hours, and ʒ ss of whiskey every two hours, with milk and eggs. He continues better for some hours. Towards evening the œdema again shows itself. He is again cupped, and gr. x ammon. carb. is ordered every two hours, alternating with the whiskey. Again the œdema clears up. In addition, a can of oxygen is ordered for the night, and the patient inhales it for 15 minutes in every hour. This relieves his dyspnœa. But, towards morning, the cyanosis and œdema again appear. The cups are applied again, and the whiskey ordered, ʒ ss every hour, alternating with the ammonia, gr. x every hour. By these measures, he is carried through the night, and in the morning is easier. Nourishment in the form of milk is still kept up. He is not allowed to sleep continuously, for, during sleep, the œdema comes on. By such fighting as this—the greatest reliance being placed on whiskey, milk, and dry cups—a patient is occasionally brought through. If, on the following day, he is still worse, the resources in the way of stimulants are not exhausted. Other forms of ammonia are used. Hypodermic injections of camphor dissolved in sweet oil, are given every three hours, in 4-grain doses. If the patient has persistent œdema and a full pulse, venesection is tried, and is invaluable when digitalis and cups no longer avail. The oxygen cannot be pushed too much, as it causes unconsciousness.

Hypodermic injections of ether to the amount of 1 or 2 drachms, sometimes bring up the pulse. Teaspoonful doses of champagne every five minutes, will help to tide a crisis. There is a limit to stimulation, of course. When ℥ss of whiskey every half hour has no effect, the patient will die.

It will be seen that no new or specific treatment can be deduced from these cases. It has become a firmly-rooted belief that quinine is a good thing to give, and in those so treated, the mortality has been somewhat diminished. The class of patients is not one upon which cardiac sedatives can be fairly tried.

Dr. Austin Flint treats pneumonia by rest in bed and the administration of half an ounce of whiskey every two hours, with milk for nourishment. When the action of the heart is weak and irregular, digitalis is administered in doses of ♏xx of the tincture three times a day.

EMPYEMA.

In all cases of empyema Dr. Alonzo Clark expects to puncture several times. His practice is to incise the skin, plunge in a trocar and canula, and after withdrawing these instruments, insert into the opening thus made a linen tent, which he fastens by its free ends to the chest by means of adhesive strips. This tent is removed every day or two to allow the pus to run out. Dr. Clark is also very partial to Dr. Wyman's method, as practiced by Dr. Bowditch, of Boston. This consists in the use of the exhausting pump, and for this purpose he likes the ordinary stomach pump. He uses an extremely small trocar, and a common exploring needle, which he regards as a good instrument for this purpose, or better still, the aspirator with a fine trocar. Before introducing the instrument he considers it well to *benumb* the part by firm pressure with the finger. In these cases he has not found it necessary to use the scalpel as when a large trocar is used. After the trocar has been withdrawn the canula is pressed in some distance, as it will not hurt the lung if it touches it. The piston is worked slowly. As the opening thus made is small it closes itself to the exclusion of air as soon as the canula is withdrawn. Dr. Clarks regards this last method as peculiarly desirable before it has been ascertained that the effusion is purulent by a previous operation. When it is known that the cavity contains pus, Dr. Clark does not pretend to choose between the two methods. He continues to draw off fluid until *oppression* is felt *at* the *sternum*, and guards against drawing off too much for fear of making a vacuum, too great for the comfort of the patient. When the operation is performed in front it is done between the sixth and seventh rib, to avoid wounding the diaphragm. On the side he selects a spot between the seventh and eighth rib, and in the back between the eighth and ninth rib. He thinks that iodine injections for the prevention of further effusion of pus are attended with more harm than good. If injections of any sort are to be used he prefers simple warm water. Tonics he regards as more or less serviceable, and this he thinks is about all that can be done in the way of treatment. Even under the best management he expects half of these patients to die.

DIPHTHERIA.

Dr. C. E. Billington recommends the following prescriptions:

No. 1. *Iron and Glycerine Mixture.*

℞ Tinct. ferri chloridi............................ f℥j.
    Glycerinæ,
    Aquæ.. āā f℥j.   M.

Sig.—A teaspoonful of this and of No. 2, alternately, every half hour through the day.

No. 2. *Chlorate of Potassium Mixture.*

℞ Potassii chlorat.............................. ℨss.
    Glycerinæ..................................... f℥ss.
    Aquæ calcis................................... f℥ijss.   M.

Sig.—A teaspoonful of this and of No. 1, alternately, every half hour through the day.

No. 3. *Spray Mixture.*

℞ Acid. carbol................................... ♏xv.
    Aquæ calcis................................... f℥vj.   M.

Sig.—To be used with a small hand atomizer.

The patient is allowed to sleep for an hour or two at a time at night. When awake, doses of No. 1 and 2 are alternated every half hour. The throat is sprayed with No. 3 for several minutes at a time, whenever Nos. 1 and 2 are given. In spraying, the mouth is opened widely.

Where there is nasal implication the nose is thoroughly syringed out with warm or tepid salt-water, once, twice or three times a day. This syringing is done with the patient's head inclined forward, a two-ounce hard-rubber ear syringe is used.

Dr. Billington never applies any brush or swab to the throat. He sometimes throws a drachm of No. 1, with a syringe, directly against the affected surface in the throat. He does not give quinia or any other unpleasant medicine to children. He does not give alcoholic stimulants except where a child, who cannot be induced to take other nourishment, will take weak milk-punch or egg-nog.

The patient is nourished with an abundance of cold milk, given frequently, to which a little lime-water is often advantageously added. When the stage of extreme exhaustion has been reached in bad cases the juice squeezed from beek-steak is given.

ASTHMA.

Among the means by which asthmatic attacks are treated by Dr. Clark, is the inhalation of the smoke of dry stramonium leaves, or else he orders a piece of bibulous paper to be dipped into a solution of the nitrate of potassium until it becomes pretty well filled with nitre. This paper is then dried and a small piece of it is burnt, and the sufferer inhales the fumes. The most efficacious remedy for allaying the spasms, of which he knows, is the inhalation of the nitrite of amyl. This drug will stop the individual attacks, but will not cure the disease. Dr. Clark recognizes but one radical cure for asthma, and that is the iodide of potassium. One-half of his patients are cured by it.

SUB-ACUTE PLEURISY.

Dr. Clark aims, on the one hand, at subduing the inflammation, and on the other, at promoting absorption of the fluid. Bleeding from the arm he regards unnecessary, but when he is called at an early stage, which is seldom the case, he advises the use of cups, and repeats them as often as necessary He thinks that cups have great influence over recent inflammations, but that they do little good in sub-acute and chronic inflammations. As in other sub-acute inflammations, blisters are applied, and in doing this he selects three spots, one being placed on a new spot as the last has healed. He scarcely ever finds that more than three are required. Among active diuretics, he prefers potassii iodidi, gr. xxx per day. If this does not subserve, he tries the following:

℞ Potas. acet.,
    Inf. digitalis... āā ℨ ij–iv.
        Each day.
Or:

℞ Pulv. dig.,
    Pulv. sallæ mer.,
    Hydrarg. chlo. mit.. .... āā gr. x.   M.
        Et ft. pil. No. x.
    Sig.—One pill thrice daily.

He uses this until the effect of the mercury is produced, and then tries potass. iod. again. In some cases he finds that mild counter-irritants, such as the ammoniacal liniment, answer, as in nervous women, but, as a rule, he does not trust to these. Purgatives and vapor-baths, he sometimes finds useful. This treatment he thinks will suffice in ordinary subacute pleurisy, but when all these measures fail, he claims that we have but two resources—to do nothing or to use the trocar. Some physicians advise the early use of the trocar, but from fear of changing the serous effusion to pus, he does not like the practice. In some cases he is forced to use the trocar, but always uses medical means beforehand, if the patient is not rapidly sinking.

CHRONIC SUPPURATIVE INFLAMMATION OF THE MIDDLE EAR.

Dr. O. D. Pomeroy regards cleanliness as the chief item of treatment. He does not approve of the syringe as a means of cleansing the ear, but makes use of cotton wool. In some cases, however, where the amount of the discharge is very great he is compelled to employ the syringe. After the ear has been cleansed he uses astringents in the shape of nitrate of silver. To apply this astringent properly he makes use of a dropping tube which consists of a hard-rubber catheter with a soft rubber thimble upon its ring extremity. From three to six drops of the astringent solution are thrown into the ear with some little violence, then drawn out, and the process repeated several times while the dropping tube is still inserted in the ear. If the perforation in the drum membrane is small, and if it is evident that there are extensive purulent processes going on in the tympanum, Dr. Pomeroy does not hesitate to make an incision in the membrane, so that the astringent can be effectively introduced. The solution of nitrate of silver used has a strength ranging from 10 grains to the ounce, all the way to 480 grains to the ounce. This saturated solution has been employed upon several occasions without causing any pain or discomfort. The rule adopted is that the solution used should never be strong enough to give rise to pain.

Where there is evidence of purulent inflammation of the tympanic cavity without perforation, a strong solution painted upon the drum membrane has frequently been found to arrest the purulent process.

With regard to the use of other astringents the objection urged against *alum* is that it forms alum-curds which act as foreign bodies in the tympanic cavity. *Acetate* of lead, has been occasionally employed by Dr. Pomeroy in solutions of from 2 to 10 grains to the ounce. *Carbolic* acid has the additional value of being a disinfectant. The solutions employed vary in strength from 2 to 4 grains to the ounce.

Where there is a relaxed condition of the lining membrane of the tympanum the cavity is packed as full as possible with absorbent cotton.

Dr. Pomeroy does not believe in artificial drum membranes.

He recommends his patients to wear cotton in their ears until cured. The cotton is placed loosely in the ear and removed as often as it becomes in the least moist.

Granulations or polypi are treated by removing them with the forceps. Small nasal forceps have been found to answer this purpose very well. A *speculum* is avoided, if possible. but may be used to good advantage where the polypus is small. Hemorrhage from the root of the polypus is controlled by means of nitrate of silver or liquor ferri persulphatis.

After removing the polyp the completion of the cure is effected by means of cauterization with nitrate of silver. For this purpose a saturated solution of the crystals is employed. The method is to take a fine probe, wind the tip with a piece of cotton, dip it in the solution, remove the excess by wiping on blotting paper, and then apply it to every remaining portion of the polypus until it is completely whitened. The *strongest nitric acid* is sometimes used instead of the nitrate of silver, but is applied more carefully. In some instances *burnt alum* and *iodiform* have been of valuable assistance in removing granulations and bases of polypi. Sequestra are removed with the forceps or chopped away by means of a dentist's drill.

ACUTE PLEURISY.

This disease, according to Dr. Clark, is, by common consent, better controlled by the lancet than any other serous inflammation. He applies cups with scarification, to the affected side, and repeats them two or three times, after proper intervals. When the pain has subsided, blisters are applied, to overcome whatever inflammation remains. The treatment, then, adopted by him is decidedly antiphlogistic, so as to prevent abundant effusion. He believes that there is no occasion for diuretics. He makes use of diaphoretics in the latter stages. Fomentations of warm water, and the like, are sometimes applied to the affected side. Not much constitutional treatment is required. Dr. Clark claims that the membrane, in cases which recover, is left and is organized, and that contraction of side and shortness of breath will occur in every form of pleurisy, but that this does not take place until several weeks after the attack, and continues for three or four years, during which time the membrane becomes absorbed.

VOMITING IN PHTHISIS.

When caused by pressure of the enlarged bronchial glands upon the *par vagum*, in the early stage of phthisis, it is arrested, in the wards of the Roosevelt Hospital, by the application of dry cups between the scapulæ. Carbolic acid is prescribed, to remove the vomiting consequent upon the offensive taste and odor of the expectoration.

DIARRHŒA OF PHTHISIS.

When this symptom is present, the patients in Roosevelt Hospital are fed upon milk boiled with mutton suet until it is as thick as cream. A piece of suet is put in a bag and boiled in milk until the requisite consistency is obtained.

The following pill has been used with excellent results, in these cases:

    ℞   Resin. terebinth...............................gr. iij.
          Argenti nitrat.,
          Opii..āā gr. ¼   M.
          Sig.—One pill when needed.

APHTHÆ OF PHTHISIS.

The following mixture is used at the Roosevelt Hospital. The patient may either apply it with a brush or rinse his mouth with it:

    ℞   Quinæ sulph.....................................gr. j
          Olei piperis nigris..............................gtt. j.
          Aquæ...f℥ j.   M.

SORE THROAT OF PHTHISIS.

For the relief of this condition, salt-water and oil of black pepper, in spray, are used at the Roosevelt Hospital. The solution of salt is no stronger than that made by adding a teaspoonful of common salt to a pint of water. The oil of black pepper is added, in the proportion of 1 drop to the ounce.

JAUNDICE.

Dr. Clark recommends the free use of soda, and regards it as a better cholagogue than mercury. The form of soda prescribed is the carbonate, in ʒss doses, thrice daily. It is sometimes taken to advantage in Vichy water. As a tonic, in this condition, he prefers the proto-carbonate of iron, or the chocolate iron lozenges.

SPORADIC PERITONITIS.

Dr. Clark believes that under proper treatment a considerable number will recover, but that whatever is done must be done with energy, as the natural duration of the disease is *"four days."* Blood-letting, both general and local, he practices to a considerable extent in the treatment of this disease. Dr. Armstrong proposed blood-letting, followed by a full dose of opium, as the latter perpetuated the effect of the bleeding; but while he looked upon both as necessary, if he could have but one, he preferred the opium. Drs. Palmer and Child, of Vermont, treated their patients by the Armstrong method, in 1844, with success. When Dr. Clark first adopted this mode of treatment, eight recovered, the ninth died. His rule is to give as much opium as the patient can take without being narcotized, beginning with gr. ij–iv every two hours, until the symptoms of narcosis begin to show. In the case of a hospital patient, gr. iv were given, and the dose increased gr. j every hour until a gr. xii dose was taken. One objection to this plan of treatment, according to Dr. Clark, is, that it requires the attention of the physician, who should always administer the opium himself. It is not important which preparation of opium is used, but it is important that the same should be used from beginning to end. If pills are used they are freshly made up every twelve hours. Opium is given by its effects and not by quantity; these effects are *sensible contraction* of the pupils, marked reduction in the frequency of respiration, diminished frequency of pulse, gentle perspiration of skin, *itching* of the mucous membrane of the *nose*, and easy but very much protracted sleep, from which the patient can be easily aroused. The pain first disappears. Tympanites continues until inflammation is subdued. The bowels are let alone for one week longer, as they will move when inflammation subsides. The influence of opium is kept up until peristaltic action is re-established. The dose is then diminished, and when a spontaneous movement occurs, it is suspended altogether. A full dose is required at night to produce sleep. Dr. Clark has seen peritonitis from perforation cured by opium. No other mode of treatment has been successful in his hands. Strong coffee and the cold effusions are used by him as antidotes in poisoning from opium. With a fair amount of caution and these two antidotes, he has not often lost a patient. He does not know of a single death produced by opium in this disease.

TREATMENT OF THE PAINS OF LOCOMOTOR ATAXIA.

A male patient in Bellevue Hospital, æt. 45, had locomotor ataxia, and suffered from excruciating pains in his limbs. He had no double vision, but had slight nystagmus. There was marked delay in communication of sensation to the brain from the feet, five or six seconds passing after the foot had been pricked before the sensation was experienced. There was also persistent sensation, the pricking being felt for some time after it had been done. There was also inability to locate impressions correctly. He had not had any trouble with his bowels or bladder, and his sexual desire and capacity were unimpaired. There was no tendon reflex.

By the pains, he could predict accurately concerning the

weather. For the relief of the fulminating pains, he had resorted to the actual cautery, a variety of remedies—among them gelsemium, which relieved him for a time—and hypodermic injections of morphia. The hypodermic injections impaired his nutrition and produced delirium. All remedies used were discontinued, and for them the bisulphide of carbon was substituted, and, applied to the spine, gave, as the patient claimed, complete relief from all pain.

LOBULAR PNEUMONIA.

This condition, Dr. Clark treats by means of warm baths, the oiled-silk jacket, and those medicines which produce free diaphoresis.

GANGRENE OF THE LUNGS.

Dr. Clark's treatment of gangrene of the lungs is altogether sustaining, consisting of sustaining foods and sustaining medicines. The alcoholics are administered carefully, according to the state of the pulse. Quinia is given in 8 or 10-grain doses, thrice daily.

MULTIPLE ABSCESS OF THE LUNGS.

The treatment pursued is sustaining all the way througn, and, as there is certain to be hectic fever, Dr. Clark has found quinia to be of the greatest service.

CHRONIC PNEUMONIA.

The treatment here, again, is sustaining to a high degree. Counter-irritation is produced by means of iodine applied to the affected side of the chest. Some three or four spots are chosen, and each spot is painted about three times a day. The patient is, at the same time, encouraged to eat all the and the pack should be discontinued when it reaches 101½°. It is never to be continued until the thermometer indicates 99° lest the patient pass into a state of collapse. Cold may also be applied by sponging, and when the temperature is not very high this is often resorted to, each extremity being sponged separately, so as to expose the patient as little as possible. These two methods have superseded the use of the cold bath, as they are more convenient, more easily applied, disturb the patient less and give as good results. In severe cases the antipyretic action of quinine is added to that of the cold pack, a hypodermic injection of 15 grains being given at the beginning of the cold pack. In the third week of the fever some prefer to abandon the pack and trust wholly to quinine in large doses, fearing collapse or pulmonary complications.

For the nourishment of patients milk is used exclusively, and as much is allowed as the patient can take. Usually a full glass of milk is taken every hour, lime-water being added in the few cases where the milk alone seems to disagree. The proposal of Sir Wm. Jenner to substitute beef tea for milk, does not find much favor in hospital practice, but among private patients where beef tea can be properly prepared it is often employed. The demand is now so great that two of the New York restaurants prepare it daily for the use of the sick, and are able to dispose of large quantities.

In the latter weeks of the disease in all cases, and during its whole course in many asthenic patients, an important indication is to sustain the heart power. This is done by a free use of alcohol, whisky being the form most used at present. The quantity is determined by the needs of the case, but where there is necessity it is not spared, one case being allowed twenty ounces daily for three days with good result. Some use ammonia as well as whisky, but its action is considered less certain. When it is used the liq. ammon. acet. and the spts. ammon. aromat. are combined. Other symptoms are met by appropriate measures as they arise.

THE DIARRHŒA OF TYPHOID FEVER.

Prof. Roberts Bartholow recommends Tr. iodine, gtt.v., well diluted with water, in the treatment of typhoid fever Under this—one of the German so-called specific plans of treatment—he says, " with proper diet and nursing, the mortality is much diminished." The same writer, for the diarrhœa of that disease, prefers the following:

R Liq. pot. arsen., ... gtt. ii ;
   Tr. opii., ... gr. iv.
Repeat as often as required.

PELVIC EFFUSION.

The following are Dr. Maury's conclusions as to the treatment of pelvic effusions: 1. Caution and judgment are eminently demanded in the treatment of pelvic effusions; in the management of pelvic abscesses we should wait until maturation is complete, and simply assist Nature by making an incision as early as we are satisfied she has clearly indicated the point of opening. This is demanded in order to lessen the risk of a rupture into the peritoneum or bowel. 2. Inasmuch as many pelvic abscesses do not point at all, and manifest no tendency to open of their own accord, surgical means must be employed to make a way for their evacuation. 3. Generally these abscesses can be reached through the vagina, and whenever the effusion presents at the vaginal roof, so that it may be felt as a resisting body (it is not necessary that it should come down into the pelvis), it may be evacuated by the trocar. In rare cases these tumors present only in the rectum, or through the abdominal walls, and cannot be reached through the vagina. 4. Whenever we are satisfied of the existence of pus, and that ripening of the abscess has occurred, and thinning of the wall can be discovered, let us open it at once. 5. When we cannot, by physical signs alone prove the presence of pus, as is often the case, but believe it to be present from food in the stomach, the treatment pursued at Bellevue Hospital consists in *not* irritating the stomach, and in thereby avoiding the exciting cause of the discharges. With this object in view, the patient is put upon a milk diet. In most cases, the milk does not irritate the stomach. A certain portion of the casein is digested in the stomach, and the remainder is digested in the small intestine. Again, as the result of experience, it is known that chronic inflammation of the upper part of the large intestine is frequently best treated by rest, and a diet consisting principally of fat, in the form of either cream or cod-liver oil. By placing a patient on a milk diet, both indications are fulfilled. The real difficulty encountered in treating such cases is to change from the milk to some other diet, when a change became desirable or necessary. While they adhere rigidly to a milk diet, it is cured; but as soon as they return to ordinary food, the diarrhœa returns.

If complete control can be had of a patient, permanent cure can be accomplished in many cases by introducing the ordinary diet very gradually. First, stopping the milk entirely, not continuing it with other articles of food. Then very carefully regulating the quantity and quality of the food which the patient is to take. The first article used to the best advantage is meat—beef or mutton—finely cut and taken in small quantities at a time; and to this, perhaps, a small quantity of toast and tea, or an egg, is added three times a day. After a few days, perhaps, rice is added; and so, adding the most easily digested articles, the patient is gradually changed from one diet to another. If the diarrhœa returns, the patient is at once put back on a milk diet. The gradual transfer to the ordinary diet is sometimes aided by certain drugs. The mineral acids and preparations containing strychnia have been found to be the best that can be employed, taken with the meals. The following is often used:

℞ Tr. nucis vom.................................... gtt. xx.
    Acid. nitro-muriatic. dil........................ gtt. xx.   M.
Sig.—Dilute well with water, and take three times a day, with meals

In this way these cases are not only temporarily, but permanently cured.

THE INTRAVENOUS INJECTION OF AMMONIA.

Dr. Gaspar Griswold claims that the intravenous injection of ammonia is a prompt and powerful means of stimulation, acting efficiently in cases where other measures are of no avail, and that no bad effects follow its employment. He employs, for this purpose, a solution of aqua ammonia (equal parts of aqua-ammonia and water.) He does not perform intravenous injection through the skin, but, dissecting down upon the vein and exposing it, he then introduces the needle until the point is felt free in the interior of the vessel.

OPHTHALMIA NEONATORUM.

Dr. Knapp does not think it necessary to apply any caustic. If any is employed, he never uses a stronger solution of the nitrate of silver than 3 grains to the ounce. He believes in the great importance of cold applications. These applications are made night and day, and great care is taken to open the eyelids and carefully wash away the secretions. This is done every half hour. When there is proliferation of the mucous membrane, the nitrate of silver is the proper remedy.

ACUTE PRIMARY OTORRHŒA.

Dr. H. Knapp lays great stress upon the necessity for rest, in the treatment of this disease. His local treatment consists in injections of warm water into the ear; leeches behind the ear; inflation of the drum—at first, cautiously, through the catheter, then according to Politzer's method; astringent gargles; steaming of the ear; paracentesis of the drum; opening of the mastoid process; cleansing of the ear by syringing and wiping with a dentist's cotton-holder; the use of astringent injections into the ear, suited, in strength, to the copiousness of the discharge and the proliferation of the mucous membrane.

RECTAL ALIMENTATION.

Dr. A. H. Smith reaches the following conclusions:

1. That defibrinated blood is admirably adapted to sustaining nutrition by rectal alimentation.

2. That from 1 to 6 ounces can be retained, and that frequently a larger quantity can be used without very much trace of blood in the fæcal evacuations.

3. That in about one-third of the cases, it produces more or less constipation.

4. That in a small proportion of cases, constipation persists and necessitates the discontinuance of the blood.

5. That in a small percentage of cases, irritability of the bowels attends its protracted use.

6. That it is only an aid to stomach alimentation.

7. That its use is indicated in cases in which asthenia is developed by disease not involving the large intestines.

8. That in unfavorable cases it is capable of giving a favorable impulse to nutrition not obtainable from other sources.

9. That its use is entirely unattended by danger.

MITRAL AND AORTIC REGURGITATION.

Dr. Alfred Loomis gets rid of the pulmonary congestion by the application of dry cups. To improve the nutrition and stimulate the flagging heart he gives iron and digitalis in combination. Digitalin he sometimes finds to be the more efficient preparation of the latter drug. The dose is from $1/100$ to $1/60$ of a grain twice daily. Absolute freedom from mental excitement and from over-exertion is insisted upon. The diet allowed is principally albuminous. All stimulus is avoided. To relieve the portal circulation an occasional drastic purge is given—calomel being the drug generally employed.

MORPHIA VOMITING—OPIUM POISONING.

Dr. Montrose A. Pallen has derived very excellent results from the hypodermic injection of from 30 to 40 minims of the fluid extract of coffee into the epigastrium.

ACCUMULATIONS IN THE EARS.

Dr. Samuel Sexton has found syringing with a suitable instrument to be the best means for removing these accumulations, but he thinks that in cautious hands the curette and forceps are often of service. The water used in the syringe is as warm as the patient can bear. The syringe, when filled, is held in the right hand, while the operator pulls the auricle upward, backward and outward with the left hand, thus freely exposing the opening of the meatus. When through syringing, the meatus is dried with absorbent cotton and a firm pledget of wool is worn in the ear for a time.

EMESIS.

Excellent results have been obtained at the Presbyterian Hospital, from the employment of faradization in these cases, one electrode being placed in the right auriculo-maxillary fossa, and the other being rubbed over the stomach for several minutes.

ACUTE RHEUMATISM.

Dr. Wm. H. Thomson uses this prescription:

R Sol. acid, salicyl. (gr. xl–f℥j)......................... f℥ij.
    Tinct. gaultheriæ... f℥j.
    Aquæ.. f℥iv. M.
Sig.—A tablespoonful.

CIRRHOSIS OF THE LIVER.

Dr. Thomson orders gr. xx of the iodide of potassium thrice daily and half of the following mixture in the morning:

℞ Magnes. sulph... ℥ ss.
    Ferri sulph.. gr. viij.
    Acid. sulph. dil.. f℥j.
    Aquæ .... q. s. ad f℥ iv.   M.

In some cases a pill with the following constituents has had good effects.

℞ Belladon... gr. vj.
    Ex. nuc. vom... g. jss.
    Ex. colocynth... gr. xij.
    Aloës Socot.. gr. vj.   M.
    Et. in pil., No. vj. div.
    Sig.—One thrice daily.

URÆMIA.

Jaborandi has been found at Bellevue Hospital to be a very effective substitute for the old hot-air bath, acting more quickly and surely. It is given hypodermically in the shape of drachm doses of the fluid extract. This dose is repeated every other day.

CHRONIC NEPHRITIS.

Cases of this disease have been treated very successfully at Bellevue Hospital with jaborandi. The dose is a drachm of the fluid extract given every other morning, the patient being kept in bed until dinner-time, when the sweating is over. It has been found better not to give it at night, as the bed clothes become saturated with perspiration and sleep is disturbed and uncomfortable.

BRONCHO-PNEUMONIA.

Dr. Wm. H. Thomson puts his patients upon a milk diet,

with an allowance of f ℥ iij of whiskey a day, and prescribes the following:

℞   Ammon. carb................................,............... gr. lxxx.
    Mucilag.,
    Aquæ .... āā f℥j.   M.
Sig.—A tablespoonful thrice daily.

This mixture is alternated with one containing compound syrup of squill, wine of ipecacuhana, and sulphate of morphia.

ACNE ROSACEA.

The pustules are all opened by means of a wide and reasonably deep incision. The papules are then treated in the same manner, being cut until they bleed quite freely. Among the methods of removing the congestion, recommended by Dr. H. G. Piffard, are (1) poulticing; (2) holding the face in hot water—[this is accomplished as follows: The patient takes a basin of hot water, immerses his face, withdraws it and breathes, then immerses it again, and so on]—(3) covering the face with pieces of muslin kept constantly wet with water as hot as can be borne. If there is much congestion of the skin, where it is not invaded by the pustules and papules, little scarifications are made wherever it is most marked. Infiltration is reduced by the use of alkaline applications. The face is thoroughly rubbed three or four times a week with green soap. The red color and polished appearance of the skin following the use of the soap is most readily removed by the use of sulphur.

The following is the combination employed:

℞   Lac. sulphur,
    Glycerine,
    Rose water,
    Bay rum..āā q. s.

When varicose veins are present, they are destroyed either

by dividing them crosswise, or, still better, by dividing them lengthwise, throughout their entire extent, with a thin, sharp knife, and then rubbing a small amount of the persulphate of iron into them. In some cases, they are obliterated by means of a white-hot needle. The marked hypertrophy of the skin present later on in the course of the disease, is treated by the constant galvanic current, applied directly through the nose. When the thickening is excessive, the use of red-hot needles, or the excision of portions of the integument, becomes necessary.

CROUPOUS PHARYNGITIS.

Dr. Thomson prescribes a milk diet and f℥ iv of whiskey a day, with the following as a gargle:

℞ Sol. brominii, (℥j–f℥ ij)................................ f℥ ij.
    Glycerinæ... f℥ ij.
    Aquæ..................................q. s. ad f℥ iv.   M.

TYPHOID FEVER.

Dr. Alonzo Clark believes that a case of typhoid fever, of average severity, needs no medication except for the relief of symptoms.

Diarrhœa he manages by the following:

℞ Bismuth. subnit.. ℥j.
    Morphiæ sulph... gr. j.   M.
      Et in chart. No. xii div.
  Sig.—One to four a day.

Other astringents which he has found to be of service are tr. kino, tr. catechu, and decoction of blackberry root.

Cough, when it is the result of a catarrh, is treated either by the comp. tr. of benzoin, in 10-drop doses every three or four hours, or by means of the following combination:

℞ Mist. guaici... f ʒj-f ʒ ss
    Tr. balsam. tolu...................................... gtt. vj-x.   M.
  Sig.—Every two to four hours.

Occasionally, good has been accomplished by the inhalation of the vapor of warm water for an hour or two every day.

Restlessness is soothed by sponging the surface of the body with warm or cold water. In some cases, a Dover's powder is required.

When the temperature of the body runs very high, quinia is given in decided doses, or cold water is employed. In young persons, Dr. Clark finds the cold bath the most convenient and efficient means of reducing temperature. The temperature of the bath used is just 10° below the temperature of the patient's body. The patient is allowed to remain in the bath for 20 minutes. If the temperature rises again, another bath is given.

Hemorrhage from the bowels is controlled, if possible, by absolute rest and doses of the fluid extract of ergot.

In perforation of the bowels, the patient is placed thoroughly under the influence of opium.

Regarding diet, Dr. Clark believes in the steady and persevering administration of such food as can be absorbed by the stomach. Milk, beef tea, raw eggs beaten up with water and made of such consistency that they can be eaten with a spoon, and expressed beef-juice, are all of value. Where one disagrees, another is substituted. The expressed juice of beef is obtained by cooking a piece of steak so as just to crust the surfaces, and then squeezing out the juice with a lemon-squeezer. As the disease advances, the food administered is more and more sustaining. When the stomach fails to retain food, nutritious enemata are employed.

Dr. Clark considers plenty of *fresh air* as of prime importance to typhoid fever patients. He always insists that a window on the side of the room opposite the patient, be dropped a certain distance from the top, even in the winter season, and that the patient is protected from the draft by the use of a screen.

Bed Sores. Nothing has been found so effectual for their prevention and treatment as a water-bed. Where this cannot be obtained, the best substitute is the padded ring, or a rubber ring filled with air.

Tympanites. Cold compresses are applied to the abdomen, and covered with oiled silk. Where this fails, a stimulating injection is given, consisting of $\text{O}ss$ of a solution composed of Labarraque's solution, 1 part, and water, 16 parts. In other cases, from 8 to 10 drops of the spirits of turpentine are given in mucilage.

Abscesses are always opened early.

Peri-parotiditis is subdued by cold compresses or ice, locally.

The alvine discharges of Dr. Clark's patients are always disinfected at once. The way in which this is done is by placing in the bed-pan or vessel a half pint or a pint of a solution of the sulphate of iron, just before it is to be used.

The bed-clothes, as soon as they are soiled, are removed from the bed and plunged into a tub of water, which is sufficiently impregnated with carbolic acid to secure effectual disinfection.

Dr. Clark is in the habit of administering some mineral acid in all cases of typhoid fever—hydrochloric is that generally employed—but does not think much of the German treatment by iodine or calomel.

Two-thirds of his patients have been found to do better without stimulants than with them. A good general rule is the following: *When alcoholics diminish the frequency and increase the force of the pulse, they do good.* In such cases, they are given in a quantity sufficient to increase the force of the pulse, and to diminish the restlessness and suffering of the patient.

Within the last two or three years, the only precaution taken against contagion, at Bellevue Hospital, has been to disinfect the stools. This is done, generally, with sulphate of iron, which is placed in the bed-pan previous to its being used.

Commercial muriatic acid, diluted, is poured into the pan after the passage. The stools being disinfected, no further attempts at protecting the house-staff, nurses, or other patients, are employed.

The treatment at present in vogue is that of quinine and baths. This was begun four or five years ago, and has received such favor that it is quite the routine now. The quinine is given differently. Perhaps the most popular way has been 10 grains, two or three times a day, the evening dose being doubled, if the temperature rises above a particular height, say 105°. It sometimes causes gastric irritation, being given in powder form. If it is vomited, pills are tried, and finally, double doses by rectum. Quinidia was used for a short time, and it reduced temperature like quinine, but irritated the stomach more. Baths, in every shape, are used, but the sponge-bath is the form most adopted. The patient's temperature is taken; if found above a certain height, he is stripped either entirely naked, or perhaps only the upper half of the body. He is then sponged over with water, at a temperature of from 60° to 80°. If only half the body is uncovered at a time, that part is allowed to dry, and it is then covered, and the rest of the surface sponged. This process is kept up for 15 minutes. If that is insufficient to reduce the temperature, it is prolonged to half an hour. It is repeated every one, two, or three hours, according to the result obtained. At the end of the bath, a little whiskey is generally given.

The effect of the quinine on the temperature is to reduce it slightly in a considerable number of cases. Its effect on the patient is to produce nausea and vomiting in a small number of cases.

The sponge-baths are almost always pleasant to the patient, if not too frequently repeated. If given every hour, or two hours even, they seem to weary and annoy him. They certainly reduce the temperature in most of the cases. In a small number of these the reduction seems to last for many hours. Sometimes, two or three baths given in the afternoon .

and evening reduce the fever two or three degrees, and it keeps down for twelve hours. But it is not very rare that the baths are given every hour even, without producing very marked effect. The sponge-bath is a much more efficient antipyretic than quinine. The wet pack is hardly used now. In one case where it was employed, pneumonia complicated the disease. The plan of placing the patient in water at a temperature of 98°, and then gradually lowering it, has been tried a number of times, and so far no deaths can be be traced to it. But these gradually-cooled baths are uniformly annoying and depressing to the patients. They don't like them. Neither have they been proved to reduce temperature permanently any better than the sponge-baths do.

Several cases have been treated upon the Kibbe bed. Its action and effectiveness were similar to immersion in the bath-tub. It did not eliminate the fever from the disease.

The results of the treatment of typhoid fever patients at Bellevue Hospital, within the past ten years, has shown that large doses of quinia, and that the antipyretic treatment by cold baths are unnecessary, and that the employment of mineral acids and of symptomatic remedies is sufficient.

Dr. Alfred Loomis does not believe in the efficacy of cold baths or of large doses of quinia in arresting the development of typhoid fever. He maintains the temperature of the room in which the patient lies below 60°F. Frequent sponging of the body with cold water has been of service in his hands. As soon as the axillary temperature in the evening rises above 103°F. he places his patients in a bath of a temperature of from 70°–80°F., and then gradually lowers that temperature until the patient's temperature begins to fall. When the patient's temperature reaches 103° he is to be taken out and put to bed. When his patients are too weak to be put in the bath he employs the wet pack. He regards cold baths as contra-indicated by feebleness of the heart's action.

As an antipyretic he gives gr. xxx of quinia in one dose, or gr. x every half-hour until gr. xxx or xl have been administered. He is in the habit of administering an antipyretic dose of quinia when the temperature has been reduced by the baths.

He lays down the following rules with regard to the use of stimulants.

1. They are never to be administered indiscriminately—that is, they are never to be given simply because a patient has typhoid fever.

2. When there is a reasonable doubt as to the propriety of giving or withholding stimulants, it is safer to withhold them, at least until the signs which indicate their use become more marked.

3. In giving stimulants the effects of the first few doses are to be very carefully watched.

4. Stimulants are contra-indicated by dry tongue, restlessness, increasing delirium, pulse and temperature.

As diet he allows at first only milk diluted with lime-water and later, cream and the yolk of eggs in milk.

Diarrhœa as occurring early in the course of the disease he allows to go untreated. When it comes on in the third or fourth week he controls it by means of opium.

Tympanitis he relieves by the application of turpentine stupes to the abdomen.

Hemorrhage is treated by absolute rest, opium, internally, and ice bags applied over the abdomen.

Bronchitis he has found amenable to dry cupping and the internal administration of the carbonate of ammonium.

Laryngitis. A small blister is applied on either side of the angle of the jaw and the whole neck enveloped in a poultice.

Bed-sores are prevented by frequently bathing the parts with camphor. If the sores penetrate the integument they are washed with a weak solution of carbolic acid and afterwards covered with lint smeared over with vaseline.

Headache. Warm fomentations are applied to the forehead. Among the anodynes the best for the treatment of this symptom are the bromides and chloral.

Delirium is generally relieved by chloral and opium. In some cases the best results follow the use of stimulants.

During convalescence Dr. Loomis regulates the patient's diet with the utmost care, allowing no indigestible article of food.

ACUTE TRACOMA.

Cleansing and the use of cold are regarded by Dr. Knapp as the proper methods of treatment. He rarely resorts to division of the outer commissure. He believes that slitting of the cornea is done too frequently. He employs eserine to reduce tension. The solution of the drug used contains gr. iv of eserine to the f℥j of water.

LARYNGEAL PHTHISIS.

Dr. F. H. Bosworth first cleanses the parts by means of one of these solutions:

(1.) ℞ Acid. carbol. cryst.................................. ♏xij.
        Sodæ bicarb.,
        Sodæ biborat.................................. āā gr. xxiv.
        Glycerinæ.................................. f℥jss.
        Aquæ rosæ.................................. q. s. ad. f℥viij.   M.

(2.) ℞ Sodæ salicylat.................................. gr. x.
        Sodæ biborat.................................. ℈j.
        Glycerinæ.................................. f℥j.
        Aquæ rosæ.................................. q. s. ad. f℥viij.   M.

Whichever of these is employed is best administered by means of the Sass spray tubes with the compressed air apparatus, at a pressure of from 15 to 20 pounds. The tongue is protruded and held, thus lifting the epiglottis and uncovering the laryngeal cavity—the patient being directed to sound in high key A—the point of the tube is then passed beyond the crest, and, the pressure being turned on, the cavity is flooded with the spray. This is repeated several times until the parts are thoroughly cleansed. If the application causes pain a 5-10 gr. solution of morphia, rendered alkaline by the addition of sodii carb. or potassii carb., is used.

As astringents Dr. Bosworth employs gr x solutions of the sulphate of zinc; or argenti nitrat., gr. iij–v to f℥j; or zinci chloridi, gr. iij to f℥j; or tannin. et glycerinæ, fℨj to f℥j; or liq. ferri persulph., ♏xx to f℥j.

After applying some astringent, iodoform is used, in the following shape:

℞ Morphiæ .... gr. x.
  Tannin .... ʒij.
  Iodoform .... ʒvj.   M.
Et ft. in pulv.

Or else he employs the saturated solution in ether (♏xl-- f℥j.) The powder is applied by means of the powder-blower. This instrument was devised by Dr. Andrew H. Smith, of New York, and consists of a small, open-mouthed bottle, through the cork of which two tubes are passed, bent at right angles; to one tube is attached a single hand-ball; the other is fashioned to adapt it for the special application to be made. The powder being placed in the bottle, a single quick pressure on the hand-ball drives a current of air into the bottle, which stirs the powder up into a fine cloud, and drives it out, in this state of fine diffusion, through the other tube, and deposits it, in an evenly-distributed, thin layer, over the parts which are to be medicated. Dr. Bosworth does not believe in the use of the brush, sponge, or probe, as media of medication in this disease, but regards them as calculated to injure the diseased parts. He always uses the spray, or insufflator. He does not think much of the atomizer. He thinks the operation of tracheotomy a simple one, and one which might be performed much more frequently than it is, so giving the larynx rest.

NASO-PHARYNGEAL CATARRH.

Cleanliness is regarded as the chief requisite by Dr. W. F.

Duncan. The diseased mucous membrane is always cleansed before medicines are applied. The following alkaline and disinfecting solution has been found to be of service:

R Acid. carbol.. ʒjss.
  Sodii bibor.,
  Sodii bicarb..āā ʒij.
  Glycerinæ.. f℥ij.
  Aquæ.......................................q. s. ad fOij. M.

This is used in the atomizer, the post-nasal syringe, or the douche. The best method, in Dr. Duncan's opinion, is that by the post-nasal syringe. It is entered flat on the tongue, which is depressed by its nozzle; its point is then quickly introduced behind the palate, and the contents suddenly and forcibly ejected by driving home the piston. The hand-ball atomizer, when used with about 30 pounds pressure, has been found to be very efficient in dislodging mucus from the superior meatus. It is better for children than the post-pharyngeal syringe.

Medicines are applied in the form of spray, powder, or solution. The spray is regarded as the best medium. As an efficient astringent, Dr. Duncan employs gr. xv of the sulphate of zinc to the f℥j of water. If the case is a mild one, a solution is not more than one-fifth this strength. Applications are made three times a week, and in the intervals, the patient applies the cleansing solution above given, himself, by means of Delano's atomizer, or the post-pharyngeal douche. When there is excess of secretion, and but little sensibility, ferric alum (gr. v–xx to aqua f℥j) has been found to be useful. His general rule is to ring the changes on astringents until a good one has been found. Others than those already mentioned, which he employs, are chlorate of potassium, nitrate of silver, tannin, and chloride of zinc. When pain follows the use of an astringent, a spray of morphia is employed. When stronger applications are needed, caustics are applied with a probe, one end of which is tightly wrapped with cotton.

The probe employed has a short arm of an inch in length, so that applications can be made with it behind the palate to the vault. Hypertrophied tissue is always destroyed by means of forceps, knife, or galvano-cautery. Polyphoid thickening of the ends of the turbinated bones is touched with caustic applied by means of a probe passed through a shield. Where there is adenoid degeneration the vault is curetted.

In the atrophic form of naso-pharyngeal catarrh, stimulating solutions are employed by means of the spray, such as a weak solution of iodine, (gtt. v–x to aqua f℥j,) or tincture of sanguinaria, (f℥j to aqua f℥j.) Sometimes, sanguinaria, myrrh and lycopodium, in powder, is blown into the nostrils.

The simple ozæna is treated by carefully removing the pellicle every day or so, and then using an astringent spray, after which iodoform powder is blown into the nostrils. The nasal passages are kept constantly open and dead bone is removed at once.

THE LARYNGITIS OF LEPROSY.

Dr. Louis Elsberg cleanses the larynx thoroughly with the spray and then administers soothing inhalations. A diluted emulsion of garjun oil has acted very well in his hands as a local as well as an internal remedy. He has had excellent results from local applications of a solution of iodoform in sulphuric ether.

CHRONIC BRIGHT'S DISEASE.

The diet allowed at Bellevue Hospital, in this disease, consists mainly of milk, eggs, butter and fresh fish. Meat and fried fats are not allowed. Those vegetables containing the least amount of woody fibre, such as rice, potatoes and onions, are given freely. Asparagus, turnips, cabbage and beans are not desirable. For the anæmia, iron is administered in the form of the tincture of the chloride, combined with nux vomica and sweet spirits of nitre:

℞ Tc. ferri chlor.,
  Tc. nuc. vom................................ āā f℥ij.
  Spts. ether. nit.............................. f℥jss.   M.
  Sig.—A teaspoonful thrice daily.

Cod liver oil is given to increase the nutrition, and to combat the disease itself the following prescription is employed:

℞ Hydrarg. chloridi corrosivi................ gr. j.
  Ex. dig.,
  Quiniæ sulph.............................. āā gr. xx.   M.
  Et in pil. No. xx div.
  Sig.—One pill three times a day.

To let out the serum from the legs punctures are made in the skin and the legs then wrapped in cloths wet in a solution of carbolic acid in water, to which essence of cinnamon has been added. The patient is rubbed all over once a day with sweet oil.

THE MANAGEMENT OF THE DIARRHŒAS OF CHILDHOOD.

Beef Tea in Diarrhœa.—Dr. Smith has found that beef tea in the diarrhœa of children almost invariably acts as an irritant and aggravates the disease. Sometimes it seems to pass the bowels in the same form in which it was taken. In any case of acute diarrhœa he advises not to give beef tea.

The Reduction of Temperature in Diarrhœa.—The best means of reducing the temperature according to Dr. Smith is by the external application of cold. Since Kibbe's cot has been devised, the immersion of the child in a bath is practically done away with. Kibbe's cot can be improvised easily; it is a pleasant and convenient way of giving the wet pack; is just as effectual as the bath, and has very few of its objections. Fold a small sheet so that it will cover the child from the axillæ to the ankles, place the child on the bed, leaving the arms and feet uncovered. The axilla can be dried easily, and the temperature be taken while the child is in the pack, or the thermometer is introduced into the rectum, the most accurate way of taking the temperature. Water of the desired temperature is poured on from a pitcher. In cases of slight elevation of temperature, say to 102° F., or under, sponging off the body with water about the temperature of 80° F. has usually been found to answer the purpose, and it is done often enough to reduce the temperature nearly to normal. But in all cases of an elevation of temperature above 102° F. resort is had to the Kibbe's cot or its substitute. Dr. Smith always remains and makes the first application himself. The temperature of the water used is at first 90° F., then gradually, as the child becomes accustomed to it, it is made cooler, until it is brought down to 80° F. in a few minutes. It has been found to be necessary where the temperature is very high, or where it rapidly rises after it has been reduced, to apply the water even colder than 80° so as to reduce the temperature to 99°. It usually goes down still farther after the child is taken out. After removing the sheet, the child is put in a thin blanket, covered up and allowed to go to sleep. In very severe cases, where the temperature rises to 105° F., or higher, it is necessary to apply the cold every hour or two. In such cases it is not necessary to remove the child from the Kibbe's cot, but it may be allowed to remain there for days if necessary. The cot is made comfortable by folding a woolen blanket and putting it under the child.

Dysenteric Diarrhœa.—Dr. A. A. Smith has found small doses of castor oil and opium, given in mucilage, an excellent combination in this complaint, as in the following prescription:

℞ Ol. ricini.. ℨj.
    Sacch. lactis..................................... ℨss.
    Tinct. opii camph............................ ♏xxxij-fℨjss.
    Mucilag. acaciæ,
    Aquæ puræ.......................... āā q. s. ad ℥j.   M.

Sig.—One drachm every two or three hours.

The paregoric should be added according to the age of the child: for a child under a year, 4 to 8 drops; for a child of one to two years, 10 drops. The diet is, at the same time, to be carefully regulated. It has been found well, sometimes, in these case, to give starch-water enemata. If the enemata are given, the paregoric is left out of the castor-oil mixture, and laudanum is put in the enema. One or two drops of laudanum, with one to three tablespoonfuls of starch-water, are given, according to the age of the child. The starch-water is made about as thick as thin cream, and given tepid. It is repeated every three to six hours, according to the severity of the attack.

Flatulent Diarrhœa.—Dr. Smith has found the following prescription an excellent one in such cases:

 ℞ Magnes. calcin.. ʒj.
   Spts. amm. aromat..................................... ♏xl.
   Tinct. assafœt.. ʒj.
   Anisette... ʒvi.
   Aq. cinnamomi.................................q. s. ad ℥iv. M.
  Sig.— ʒj every half hour until relieved; to a child from three weeks to four months old.

Two or three doses will usually relieve.

Koumyss in Diarrhœa Dependent Upon Non-Digestion of Sugar.—Dr. Smith has found this to be a combination that is easily assimilated. The koumyss is charged with carbonic acid gas, but children do not take it readily with the gas in. It is gotten rid of by taking the koumyss out of the bottle and pouring it from one pitcher to another a few times. A small quantity is kept out for immediate use, and the remainder put back into the bottle and the bottle corked and put in a cool place. Sometimes children who are unable to retain anything else, have been found to take a teaspoonful of koumyss at a time and digest it, and frequently, without any medicinal treatment, will recover under its use. Dr. Smith has found that twelve hours is as long as it can be kept safely, after once uncorking it. The child need take no other food while it is taking the koumyss. It is itself food and drink. It is sour, and mothers are tempted to sweeten it to make it palatable. It is never to be sweetened, and never to be given within two hours after any other form of milk, and must be given cold. After the first repugnance to it, children take it quite readily; even children as young as six or eight months can be made to take it by taking advantage of their thirst and giving it at first in small quantities. Koumyss may be used in many forms of diarrhœa because of its easy digestion. That made by Dr. E. F. Brush, of New York city, is the only preparation of it which Dr. Smith has found reliable.

Diarrhœa Due to Inflammatory Disorders.—The indications according to Dr. Smith, are to reduce the temperature, regulate the diet, surround the child by the best possible hygiene, put the warm applications over the abdomen, and give internally a combination of opium and camphor. Tully's powder, which consists of morphine, camphor, and prepared chalk, has been found to make a good combination. The dose for an adult is the same as Dover's powder. Ten grains contain one-sixth of a grain of morphine and a little over three grains of camphor. A child three to six months old is given an eighth of a grain every two to six hours, according to the severity of the attack and the control the powder has over it. A child six to eighteen months is given one-sixth to one-fourth of a grain in the same way. After the acute symptoms have been controlled there remains in many cases a tendency to looseness of the bowels, with very little constitutional disturbance. The Tully's powder is then stopped and the following given:

℞   Ac. sulph. dil... ℳ xxiv.
     Salicin... gr. xxiv.
     Glycerinæ... ℥ iij.   M.
Sig.— ʒ i. t. i. d.

This should not be given within a half-hour of the taking of milk. The sulphuric acid has a tonic and astringent effect, and the salicin, besides its tonic effect, acts also as an anti-fermentative.

morbid material in the blood than at the other periods named.

Please bear in mind these rules which I have just given you, for you will find that they will stand you in good stead in all these cases of obstinate malarial attacks.—*Roberts Bartholow.*

LUMBAGO.

According to Dr. Alfred Stillè, the best treatment in acute lumbago, at first, is the application of wet-cups to the muscle or muscles affected, to be followed immediately by narcotic fomentations in the shape of a bag of hops soaked in hot water, hot vinegar, or alcohol, applied directly over the scarified parts. There are various stimulating and anodyne liniments which he also uses such as turpentine, ammonia, and camphor. Opium in the form of a ten-grain Dover's powder, given early, relieves pain and produces diaphoresis. Atropia hypodermically (one eightieth of a grain) is valuable, but must not be given to nursing women. Morphia may also be given hypodermically (except in pregnancy) and these two remedies are usually the best in private practice when wet-cups cannot be used. Iodide of potassium, in doses of five to ten grains every three hours, gives very good results. The most useful class of remedies in chronic lumbago are blisters, sinapisms, the actual cautery, etc. Local friction and *massage* conscientiously applied are often useful when counter-irritants fail. Tepid water may be applied, either in the shape of wet compresses kept in constant contact with the part, or in the form of a douche falling steadily upon the rheumatic muscles for some time from a height of eight to ten feet. The action of water, though slow, is a very permanent one. After the treatment by douche or by wet compresses the parts should be briskly rubbed with a coarse cloth or a skin-brush, and then covered with cotton or wool or a piece of India-rubber cloth. The use of a metallic brush is sometimes advantage ous, and finally tying the cloth over the lumbar regions and ironing them thoroughly two or three times every day, following this up with the application of some stimulating liniment, is often to be advised.

CATARRH OF THE STOMACH.

Acute Catarrh.—The most essential point in the treatment is a proper regulation of the diet. The most uniformly applicable food is milk. Care must always be taken to secure milk of good quality.

When there is considerable nausea and frequent vomiting, from two to three fluid ounces of milk, with one fluid ounce of lime-water, every two hours, will usually be retained without difficulty. Such a plan, supposing the feeding to be discontinued through the night (which, by the way, should always be done, unless otherwise demanded by excessive prostration, if there is a tendency to sleep,) supplies about sixteen or twenty-four fluid ounces of milk per diem. If the irritability is too great for the retention of these doses, the same combination of milk and lime-water must be given in diminishing amounts at correspondingly short intervals, until the proper measure, if it be only a teaspoonful, is reached. On the contrary, when the vomiting is less obstinate, or as the patient improves, the doses and intervals should be increased, and as opportunity offers, the diet gradually extended, by the addition of farinaceous articles, broths, chicken, etc., until the ordinary food is resumed. Very exceptionally, milk and lime-water cannot be retained. When this happens, it is best to abandon the milk diet altogether and substitute carefully-prepared beef-tea or chicken or mutton-broth, entirely free from fat, in doses of two fluid ounces every three hours. Beef-juice is also serviceable under these circumstances. Dr. Starr has never had occasion to use either whey or artificially-digested milk, each of which is highly recommended in obstinate vomiting. Thirst, which is often a distressing symptom, is

℞ Sodii bromid.. ʒss.
   Mucilag. acaciæ,
   Aquæ puræ..............................āā q. s. ad ℥ij. M.
   Sig.—A teaspoonful every three hours.

The bromide diminishes the reflex disturbance, and the mucilage is soothing to the irritated intestinal mucous membrane.

General Hints.—Whatever the cause, Dr. Smith insists that all children, whether infants or those older, be kept quiet when suffering from diarrhœa. They should be kept, according to his thinking, in a partially darkened, quiet room, free from noise, and all talk in the room should be avoided, especially when the child is asleep. The nervous system in childhood is so impressible that it is easily disturbed, and any disturbance of this character aggravates the diarrhœa. Infants under one year are kept lying down as much as possible. They are not jolted up and down as is the custom of most nurses and some mothers, in order to amuse them. If the child is under one year of age it is placed on a pillow, if the diarrhœa is severe, as it can be kept quiet more easily in this way than when lying on the lap. Even in changing the napkin care is taken to move the child as little as possible. The room in which the child lies is kept well ventilated. Mothers usually are over-careful for fear the child may take cold, and on this account are apt to keep the room too closely shut up. When the child is awake it is carried carefully into open air, always in the shade. Dr. Smith believes salt-air to be beneficial to almost all forms of diarrhœa in children, and especially as occurring in city children. In all cases, in children under a year, if the diarrhœa is severe, warm applications are applied over the abdomen in the shape of a spice bag. To make this take a half ounce each of cloves, allspice, cinnamon, and anise seeds pounded, but not powdered, in a mortar, put these between two layers of coarse flannel, about six inches square, and quilt them in. Soak this for a few minutes in hot spirits (brandy, or whiskey, or alcohol), and water equal parts, and apply it to the abdomen warm, renewing it when it gets cool. In this way it is possible, not only to get the effects of a poultice, but also the sedative and antiseptic effects of the spices. Great heat, with influences that depress the nervous system, bad hygienic surroundings, improper diet, too early weaning, bottle food, and dentition, are among the causes that predispose to diarrhœa. In all cases these must be removed.

Cholera Infantum.—The two special indications are to reduce the temperature and control the nervous manifestations. Cold applications are made. Hypodermic injections of quinine and morphine are given: To a child of six months, 1 grain of quinine and about $1/200$ of a grain of morphine every four or six hours, according to the indication; for each additional six months of age, an additional ½ grain of quinine and an additional $1/200$ of a grain of morphine. These are the solutions of quinia and morphia which Dr. Smith uses:

R Morph. sulph... gr. ss.
    Aquæ destillat.......................... .... ʒj.   M.
Sig.—Five minims, by hypodermic injection, for a child six months old.

R Quiniæ sulph... ʒj.
    Ac. sulph. dil... q. s.
    Acid carbol. cryst... gr. v.
    Aquæ destillat.. ʒj.   M.
Sig.—Eight minims, by hypodermic injection, for a child six months old.

Usually, the stomach is so irritable that medicines and food are both vomited. After the temperature is reduced, and the nervous system is rested, small quantities of food are given. Small pieces of ice are also given to allay thirst.

The indications for treatment in the exhaustive form of the disease are thought by Dr. Smith to consist in checking the enormous loss of fluid and in sustaining the patient. His main reliance is on opium and alkalies and stimulants. Opium, in small doses, in addition to the other effects claimed for it, he regards as a cardiac stimulant, thus meeting one of the chief indications in this disease.

The following combination is good:

 ℞ Tinc. opii. camph.. ℨiij.
   Mist. cretæ..: ℥iij. M.
 Sig.—A teaspoonful every second or third hour to a child of six months.

Sometimes nothing is retained by the stomach. In such cases, it is necessary to give the opium hypodermically in $1/200$-grain doses, without the quinia.

Alcoholic stimulants are given, and among them brandy is considered to be best. Dr. Smith gives 5 drops of brandy in a teaspoonful of water, every hour, to a child of six months, if there is great exhaustion. This quantity is increased or diminished, according to the indications. In some cases of cholera infantum, a child becomes suddenly much more exhausted, pulse becomes more rapid, extremities are cold, perspiration comes out freely, and the child seems to be going into collapse. An enema of hot water sometimes revives such a child wonderfully. A good quantity of hot water must be used, say half a pint, and a towel must be held to the anus afterward, in order to have the water retained as long as possible. Along with this, spirits of camphor are given internally, in from 6 to 10-drop doses. It may be put in with the brandy, and the two given together for a few hours. In any case of diarrhœa, where these symptoms of great exhaustion occur with the coldness of the extremities, the hot water enemata may be given.

TYPHOID FEVER IN CHILDREN.

Dr. A. Jacobi considers typhoid fever in childhood as a very manageable disease, if taken in time.

High temperature he reduces by means of the cold bath, and if reaction is not at once established he plunges the child into a hot bath, which has the effect of restoring the circulation on the surface of the body and thus enabling the blood to throw off a part of its heat. Where the child will not bear a cold bath, he bathes the surface of the body with cold water as far down as the thighs.

He regards *cold packing* of the trunk and abdomen as of great service. The child is retained in the pack until both pack and surface become slightly warm.

When temperature is high he gives quinia—from 8 to 15 grains daily in one or two doses. If the temperature is very high, from 12 to 16 grains are given at a dose. He always administers it in solution and in the form of the muriate, or neutral tannate.

He does not hesitate to recommend salicylic acid and salicylate of sodium in the reduction of high temperatures. He gives salicylate of sodium to a child in doses of 3 to 6 grains, three, four, or five times in the course of twenty-four hours. He has sometimes succeeded in reducing temperature by a combination of quinia and salicylic acid where both remedies failed to do so when employed separately.

Now and then he uses digitalis, in small doses, to invigorate the heart's action. Veratrum viride he does not recommend, by reason of its occasional irritant effects upon the mucous membrane of the stomach and intestines.

He believes stimulants to be especially indicated, and gives a baby one year old an ounce of brandy or whiskey in the course of twenty-four hours. He always gives them in milk, water, or barley-water, and never alone.

Camphor he regards as an excellent stimulant, and gives from 2 to 10 grains of camphor, in the course of a day, to a child from two to four years of age. Musk, also, he regards as a valuable stimulant. The dose for a child two years old, is 2 grains, to be repeated every hour. With it he has obtained admirable results, when nothing else seemed to be of any avail whatsoever. When a speedy effect is desired, he does not rely alone upon the internal use of stimulants, but proceeds at once to inject, hypodermically, ether, brandy, alcohol, or camphor dissolved in ether, oil or brandy. He is very careful that the solution of camphor thus obtained is not too strong, in which case the menstruum is absorbed very rapidly and the camphor is left remaining in the subcutaneous tissue.

Hemorrhage from the bowels Dr. Jacobi treats by opium and alum. In some cases he has had excellent results from the steady application of an ice-bag over the ileo-cæcal valve.

HEADACHE.

Nervous Headache.—Dr. A. A. Smith has found this form will very often yield to a saline cathartic. In addition to the cathartic, he uses this formula:

℞ Sodii bromid... ʒ vj.
    Elix. valer. ammon..................................... f℥ iv.   M.
    Sig.—A teaspoonful every hour until relieved.

When the headache is associated with anæmia, he gives iron in addition to the above, and stimulates the heart with:

℞ Ammon. muriat.. ℥ ss.
    Tc. actææ racemos.,
    Aquæ .... āā f℥ iij.   M.
    Sig.—A dessertspoonful, after meals, in a wineglassful of water.

Where there is despondency and depression of spirits, he frequently uses, thrice daily, a pill composed of gr. $1/50$ of phosphorus, and gr. $1/8$ of nux vomica.

If there be persistent sleeplessness, he gives:

℞ Camph. pulv... gr. xxv.
    Ext. cannab. ind.. gr. x.
    Ext. hyoscyami.. gr. xx.   M.
    Et. in pil. No. x div.
    Sig.—One at night; repeat in two hours, if necessary, to produce sleep.

Sick Headache.—Good results have been obtained from the use of guarana or paullinia sorbillis. The latter is administered in powder, 15 grains being given every fifteen minutes until six doses have been taken. It has been found best to give it in a little sweetened water.

Malarial Headache.—The treatment is begun by administering 15 grains of quinia every two hours before the expected attack. If quinia fails to ward off the attacks, Fowler's solution and tincture of belladonna are given in 5-drop doses.

Headache Dependent on Gout.—This is combated by means of the following:

R   Vin. colch. sem... f℥ iij.
    Lithii brom.,
    Syr. zingiber..................................āā f℥ ss.
    Aq. cinnamom........................q. s. ad f℥ vj.   M.

SIG.—A tablespoonful in a tumblerful of Vichy water every few hours.

The Headache of Syphilis.—Calomel is given in doses of the $1/10$ of a grain every hour during the day. This treatment is continued for two or three days and then stopped, and iodide of potassium given in 15-grain doses at meals. This amount is gradually increased until iodism is produced.

The Headache of Rheumatism.—He uses the mild faradic current to the scalp, and internally the following:

R   Potas. iod.,
    Ammon. muriat..........................āā ʒ iss.
    Infus. humuli.................................. f℥ vj.   M.

SIG.—A tablespoonful four times a day, in a wineglassful of water.

In cases which do not yield to this treatment, 20-grain doses of the bromide of ammonium have been found effectual.

Uræmic Headache.—Dry cups are applied over the region of the kidneys, and the following given internally:

R   Potas. acet...................................... ʒ vj.
    Infus. digital...................................... ʒ vj.   M.

SIG.—A tablespoonful every three hours.

The infusion is made from fresh English leaves. This mixture is administered until the kidneys act freely. If it does not relieve the headache within 24 hours, a saline cathartic is given. A domestic remedy often employed consists of an ounce of cream-a-tartar in a quart of water. The whole of this is taken in the course of 10 hours, and acts both as a cathartic and diuretic.

The Headache of Cerebral Effusion.—If convulsions seem to be imminent, the kidneys are stimulated by means of the digitalis and acetate of potassium mixture, given above, and from 12 to 20 ounces of blood are taken from the region of the kidneys, by means of wet cups.

The Headache of Acute Alcoholism.—To remove the alcohol from the intestinal canal, give ½ drachm each of rhubarb and calcined magnesia, and then administer:

℞   Spts. ammon. aromat............................ f℥ ij.
     Tc. camph.. f℥ jss.
     Tc. hyoscy... f℥ ijss.
     Spts. lav. comp............................. q. s. ad f℥ ij.   M.
Sig.—A teaspoonful every hour, until the headache is removed.

Gr. ij of capsicum and gr. iij of quinia are then given before each meal, for several days.

If there is sleeplessness, the following formula can be used:

℞   Sodii bromid...................................... ℥ ss.
     Chloral hydrat................................... ℨ ijss.
     Syr. aur. cort.................................... f℥ ss.
     Aquæ... f℥ iijss.   M.
Sig.—A tablespoonful at night. To be repeated in two hours, if necessary, to produce sleep.

Dyspeptic Headache.—If there is indigestible food in the stomach, an emetic, such as mustard and warm water, or gr. xv of the sulphate of zinc, is given. If the headache is frontal, 10-drop doses of nitro-muriatic acid, well diluted, are given after meals. If the pain is located about the roots of the hair, gr. xx of the bicarbonate of sodium or magnesium, are given before meals.

When the pain spreads over the entire head, Dr. Smith gives:

℞ Sod. bicarb.. ℨijss.
    Ac. nitro-mur. dil.. f℥ij.
    Tc. nuc. vom .... f℥jss.
    Syr. aurant. cort.. f℥vj.
    Aquæ..q. s. ad f℥vj.   M.

SIG.—A tablespoonful, after meals, in a wineglassful of water.

If there is gastric pain, a mustard-plaster is applied to the epigastrium. If flatulence is troublesome, bismuth and nux vomica are given combined:

℞ Bismuth. subcarb....................................... ℨjss.
    Tc. nuc. vom................................... .... f℥jss.
    Tc. card. co.,
    Spts. lav. comp............................āā q. s. ad f℥iv.   M

SIG.—Two teaspoonfuls, before meals, in a wineglassful of water

If there is constipation, the following pill is given:

℞ Aloës pulv .... ℨss.
    Ex. nuc. vom.. gr. v.
    Ex. belladon... gr. iv.   M.
Et. in pil. No. xv div.

In other forms of headache associated with indigestion, a single drop of the tincture of nux vomica is given every fifteen minutes until 10 or 15 drops have been taken.

Where headache depends on delayed stomachic digestion, ½ a drachm of saccharated pepsin, in a wineglassful of sherry wine, is effectual.

The Headache of Acute Cerebral Congestion.—The patient is kept in a darkened room, perfectly quiet, and cold and evaporating lotions are applied to the head. A saline cathartic is first administered, and then the following:

℞ Sodii bromid... ℨijss.
    Fl. ex. ergot... f℥ijss.
    Syr. zingiber.. f℥ss.
    Aq. aurant. flor.................................q. s. ad f℥iv.   M

SIG.—A tablespoonful every two hours.

If the skin is hot and dry, and the pulse full and rapid, 2 drops of Fleming's tincture of the root of aconite are given every two hours until the heart's action is sensibly reduced.

The Headache of Passive Cerebral Congestion.—The heart's action is stimulated by the use of the following:

 ℞ Tc. digital... f℥ij.
  Spts. ammo. aromat............................... f℥vj.
  Spts. lav. comp.,
  Syr. simp.....................................āā q. s. ad f℥ iij. M.
 Sig.—A teaspoonful every four hours.

The Headache of Cerebral Anæmia.—The patient is allowed to inhale from 3 to 5 drops of the nitrite of amyl, placed on a piece of cotton and applied to one nostril while the other is closed.

When associated with nervous exhaustion, this formula is used:

 ℞ Strych. sulph... gr. ss.
  Tc. ferri chlor... f℥ij.
  Glycer... f℥ss.
  Infus. gent.......................................q. s. ad f℥vj. M.
 Sig.—A tablespoonful, after meals, in a wineglassful of water.

A tablespoonful of brandy after each meal, and a glass of champagne at dinner, are often of great advantage in these cases.

The Headache of Cerebral Tumors.—The iodide of potassium is the indication followed.

The Headache of Cerebral Softening.—This is palliated by opium and rest. Ergot in large doses—f℥j–f℥iv of the fluid extract thrice daily—has been employed in several cases of this kind, by Dr. Smith, with admirable effect.

CATARRHAL INFLAMMATION OF THE MIDDLE EAR.

Dr. Samuel Sexton treats these cases by means of an instrument consisting of a soft rubber bulb, connected by a flexible tube with a hard rubber nozzle which is somewhat olive-shaped and adapted to fit into the entrance of the external auditory meatus. When used, the nozzle of the instrument is held to the ear with the left hand of the operator and pressed against the opening with sufficient force to make the fitting as nearly air-tight as possible; while the bulb is held in the right hand, to be manipulated as desired. Care is always had in using this instrument not to employ two much force in stimulating the membrana tympani. When the instrument is used for suction, so as to draw the membrane out as far as possible, the ball is simply collapsed by pressure of the right hand before the nozzle is applied to the ear. Upon removing the pressure of the hand, the bulb, by its own elasticity, gradually resumes its expanded condition, thus rarefying the air in the external meatus. If it is desired to give the membrana tympani and ossicula motion, the bulb is alternately compressed and expanded while the nozzle is in contact with the ear. Where inflammation is high, leeches are applied and warm water instillations are practiced.

CHRONIC BRIGHT'S DISEASE.

Milk is used freely. The diet is rendered non-nitrogenous as far as possible. Cod-liver oil is given in combination with the following mixture:

R Tr. ferri muriat.,
   Tr. nucis vom.. āā gtt. x.
   Spts. etheris nitrosi................................... f℥j.   M.
SIG.—To be taken thrice daily.

If the patient is troubled with gastritis, the iron and cod liver oil are stopped and cream of tartar administered as a purge. Large doses of pepsin and of the oxalate of cerium are used to quiet the stomach. The specific treatment of this disease at Bellevue Hospital consists of the following:

R   Hydrarg. bichlor.. gr. 1/16.
    Dig.,
    Quin. sulph.. āā gr. j.   M.
    SIG.—To be given thrice daily.

The skin is freely rubbed with olive oil twice daily.

CATARRHAL LARYNGITIS.

To break up the tendency to a continuation of the disease, after recovery from the attack, the child is given the following mixture, at Bellevue Hospital:

R   Hydrarg. chlor. mit.,
    Quiniæ sulph ....āā gr. j.   M.
    SIG.—To be taken three times a day until 6 grains of the calomel have been administered.

THE GENERAL TREATMENT OF CONSTIPATION

Constipation Dependent Upon Deficient Secretion.—Dr. Thomson believes that the use of cathartics or of medicines calculated to stimulate nerve action, only does harm in this condition. The best results have followed the use of large quantities of water. His patients are directed to take two tumblerfuls of water upon rising. A small amount of common salt may be added to this water to increase its laxative effect. He has found that the addition of small quantities of quinia to salines increases their power; for instance, he gives

R   Mag. sulph... ʒj;
    Quin. sulph .... gr. j,

in a tumblerful of water every morning. He does not place much confidence in the value of fruits as laxatives.

Flatulence is overcome by the following:

R   Assafœt... gr. iv.
    Saponis... gr. ix.   M.
    SIG.—To be taken when necessary.

Constipation Due to Want of Peristaltic Action.—A sitz-bath is ordered every night, the water used being as cold as the patient can bear; or, upon rising in the morning, the spine and abdomen are sponged with cold water. In other cases, great benefit is derived from dashing water against the abdomen while the patient stands up. Nux vomica, combined with soap and rhubarb, has proved very serviceable. The application of the faradic current, one pole of the battery being placed over the spine, and the other passed up and down over the abdominal walls, has often been of great benefit. In still other cases, the health-lift is recommended.

Constipation Due to Chronic Inflammation of the Rectum.—The rectum is emptied by means of enemata. When it is thoroughly cleansed out, strychnia is injected locally, a fold of the mucous membrane being drawn down and the needle inserted. This same hypodermic method has given the most excellent results in Dr. Thomson's hands in cases of *constipation accompanying enlarged prostate*. In every instance the rectum is first thoroughly emptied by means of an enema. Kissengen water is prescribed in the morning, and a suppository of belladonna, or stramonium, used at night. Faradization along the course of the colon, and hip-baths, are also of service.

Constipation Following Febrile Diseases is treated by means of the compound jalap powder.

Constipation Associated with Chlorosis.—Dry heat is applied to the feet and hands. In addition to the dry heat, the feet and arms are wrapped in cloths dipped in a solution of capsicum. The same application is made over the abdomen. After these measures have been continued for some time, cathartics are administered, the one most generally employed at Bellevue Hospital being the compound rhubarb pill. Of these, three are given at night, twice a week, or given every night until the bowels have been rendered soluble. Iron is not used until the bowels have been rendered soluble. The form in which it is most frequently prescribed is the following:

℞ Potas. bicarb... ℥ss–ℨij.
Ferri sulph.,
Ex. nucis vom.. āā gr. x.   M.
Et in pil. No. xx div.
Sig.—One after each meal.

To restore tone to the muscular coat of the intestines, belladonna and nux vomica are thus combined with colocynth:

℞ Ext. belladon... gr. v.
Ext. nucis vom.. gr. x.
Ext. colocynth. co..................................... ℨj.   M.
Et in pil. No. xx div.
Sig.—One at bedtime.

If the colocynth causes griping, it is prevented by the addition of ℨij of the bicarbonate of sodium and the division of the above mass into 40 pills, the dose being doubled, *i. e.*, two pills are to be administered to the patient at bedtime, instead of one.

FAVUS.

Dr. H. G. Piffard removes all the superficial crusts by scraping them off or digging them out with a pen-knife or small spatula. He first, however, loosens them by means of poultices or frictions with oil. After the crusts are all removed, the parts are smeared with sulphur or turpeth ointment (3 to 5 per cent.) The hairs are then pulled out, one by one, by means of a properly constructed forceps. After the hairs have been pulled out, a solution of the bichloride of mercury (2 or 3 grains to the ounce) is thoroughly rubbed in.

DISEASES OF THE HEART.

Valvular Lesions.—Dr. Flint advises this class of patients not to overtax the heart, allowing *only just such* an amount of physical exertion as can be done with entire comfort. He lays down the same rule with regard to mental excitement. When there is co-existent dilatation of the left ventricle and consequent dropsy, hydrogogues and diuretics are prescribed. To relieve dyspnœa, opiates and ether are administered. When the action of the heart is feeble and irregular, from 10 to 15 drops of the tincture of digitalis are given at short intervals. The chalybeates are indicated if anæmia be present.

Aortic Lesions.—He uses digitalis with a certain amount of reserve in the treatment of aortic lesions.

Angina Pectoris.—He treats the paroxysms by the free exhibition of alcoholic or ethereal stimulants—either brandy or Hoffman's anodyne; to this a few drops of laudanum are added with advantage. Nitrite of amyl is also inhaled and generally stops the spasms promptly.

CONTUSIONS.

This is the method of treatment employed at Bellevue Hospital. The parts are fomented continually with simple hot water until the pain ceases, and the contusion is then kept wet with the following lotion:

℞   Ammon. muriat.. ℥ij.
    Aceti,
    Aquæ...ää f℥ij.   M.

SCARLET FEVER.

The diet upon which all Dr. Thomson's patients are placed, consists of milk, either alone or with lime-water. In a certain class of cases he employs the carbonate of ammonium freely, but his general method of treatment until quite lately has been by the chlorate of potassium and dilute muriatic acid (gr. lxxx of the former and f℥ij of the latter, during the course of the 24 hours, diluted with Oij of water.) Now he substitutes the bromide of potassium for the chloride. The solution employed consists of a saturated solution of the bromide of potassium in water, to f ℥ ij of which ℥ j of bromine is added very slowly, the bottle being shaken constantly while the combination is being made. Dr. Thomson has found it better to add half of the quantity of the bromine first, and then to let the bottle stand for an hour or more before the remainder is added. When the bromine is dissolved in this manner, the bottle is filled with water until a 4-ounce mixture is made. For internal administration f ʒ j of the solution to f ℥ j of water is used, and of this a teaspoonful is given in a tablespoonful of sweetened water, p. r. n. The solution is kept in a dark place. Equal parts of this solution and of glycerine are used as a local application.

He advises occasional purges composed of gr. iij of calomel and gr. v of jalap. Rapid rise of temperature to 104°–106° he meets by the douche of ice-water to the head or by the cold pack. He does not recommend the cold bath. His way of administering the wet pack is this—he takes a sheet, wrings it from water at the ordinary temperature, wraps the child in it and over that lays one wrung from ice-water.

He recommends that from the very inception of the disease the body be oiled over three times a day; this relieves the itching and reduces the temperature. To relieve nephritis he employs the hot water pack, dry cups, counter-irritation over the kidneys, digitalis and warm water injections.

RUM STOMACH.

Dr. Alfred Loomis prescribes equal parts of the compound tincture of gentian and of columbo with gtt. v–xv of the tincture of nux vomica, before meals. He also employs occasional aloëtic and mercurial purges.

EMPHYSEMA.

Dr. Francis Delafield prescribes, when iodide of potassium

and all other remedies fail, gtt. xxv of aromatic sulphuric acid four times a day.

DIABETES INSIPIDUS IN CHILDREN.

Dr. Delafield administers the fluid extract of belladonna in small doses, or in some cases uses gtt. v of a gr. j to the f℥j solution of atropia once a day. If well borne the dose is increased to 5 drops twice a day, and so on until gtt. vij are given every three hours. At the same time a teaspoonful of cod-liver oil is given thrice daily, or if it disagrees with the stomach, the oil is well rubbed into the skin once a day. As a hygienic measure the child is thoroughly washed every day with warm water and soap, and is clothed in flannels which cover the whole body.

EPISTAXIS.

Ten-drop doses of Squibb's fluid extract of ergot are given after meals with the same amount of the muriate of iron. This treatment has been followed by excellent results in the wards of Bellevue Hospital.

DISINFECTANT AND ANTISEPTIC COUGH MIXTURE.

This is very much used at Bellevue Hospital.

℞ Potas. iod... ʒj.
   Acid. nit. dil... f℥iij.
   Tr. belladon... fʒj.
   Acid. salicyl... ʒj.
   Aq. camph........................... q. s. ad f℥iij.   M.
Sig.—Two teaspoonfuls in water three or four times daily.

SURGICAL AND VENEREAL DISEASES.

FISSURE OF THE RECTUM.

In young subjects, Dr. Erskine Mason keeps the bowels in a soluble condition, and applies, locally, zinc or stramonium ointment, in combination with belladonna or opium; or he pencils the fissure to its bottom with a fine point of nitrate of silver, or with nitric acid.

His radical method of treating this painful affection, consists in dividing the mucous membrane and some of the fibres of the sphincter muscle, with a knife, being careful to divide the nerve filament involved. In some cases, he is able to effect a cure simply by over-distension of the rectum. The thumbs are introduced, back to back, into the rectum, when forcible distension is made towards the tuber ischii, and carried to the fullest extent possible. The cutting operation is thus performed: the parts are put upon the stretch by introducing a speculum, and when they are moderately tense, the knife is drawn through the base of the fissure. As after-treatment, the patient is kept in bed for a day or so, and the bowels are quieted by means of an opiate. Before the bowels are allowed to be moved, an enema of sweet oil is given.

LUXATION AT THE HIP JOINT.

The patient is placed on his back, on the floor; the femur is flexed upon the abdomen, until it is brought at right angles with the pelvis; then, standing astride of the patient, the assistant (Bellevue Hospital) clasps his hands under the legs, close up to the thighs, and suspends the body. When the patient's body has been raised free from the floor, the sound limb is so balanced against the leg of the surgeon that the entire weight of the patient's body can be utilized as an extending force upon the dislocated limb, and assisted, perhaps, by a trifling rotation, will draw the acetabulum over the head of the bone.

NÆVUS.

Dr. R. W. Taylor employs the electric needle, for the purpose of destroying these growths. Dr. A. C. Post believes that if there is no tendency to spread, the nævi had better be let alone, but that, when they manifest a tendency to spread, they should be at once removed. In this operation, he employs an instrument consisting of six needles, each about the size of a fine knitting-needle, and without points. These needles are inserted in a handle, and, about half an inch from their outer extremities, are passed through a bar containing two rows of holes. Each row contains three needles, and the perforations are separated nearly one-fourth of an inch in each direction. By using this instrument, the operation is greatly facilitated, and the heat is better retained in this bundle than it would be in a single needle.

POTTS' DISEASE OF THE SPINE.

The plaster-of-paris dressing is thus applied at Bellevue Hospital: the patient puts on a light wrapper, and then, by means of a pulley attached to straps which pass through the axillæ and are secured to a cross-bar over the head, is completely suspended. In this way the weight of the body is made to act as a counter-extending force, and removes all pressure from the diseased surfaces of the bones. While thus suspended, the wrapper is drawn down smoothly over the trunk and hips. Then with a roller bandage filled with plaster a beginning is made 2 or 3 inches below the crests of the ilii, sufficient to secure a firm hold upon the pelvis. Two or three extra turns of the roller are first made at this point and the body is then embraced by successive turns until the entire trunk is encased to the arm-pits. If the disease is high up in the dorsal region, the bandage is carried over each shoulder, like a pair of suspenders. A roll of cotton is placed upon each side of the spinal cord opposite the seat of disease. Two or three rollers, supported by fine strips (made rough like a grater to prevent them from slipping,) make a substantial jacket. The cotton padding is not allowed to be excessive, as it would thus obliterate the irregularities of the surface almost entirely, which is just the thing to be avoided, for it is the gentle moulding of the jacket to the depressions and elevations of the surface of the body that enables it to accomplish its work and be worn without complaint. When abscesses are present, windows are made in the jacket through which they can discharge.

STRICTURE.

Dr. H. B. Sands believes internal urethrotomy to be applicable chiefly to close strictures and as an auxiliary to dilatation. He is a firm believer in gradual dilatation as by far the best method of treating the majority of strictures. He considers the use of sounds exceeding 25 mm. to be very rarely necessary either as a means of diagnosis or of treatment. The practice of slitting up the meatus he denounces as irrational.

Dr. F. N. Otis regards gradual dilatation as only a temporary expedient. He does not think that section should be limited to close strictures. In all cases he insists upon the prompt restoration of the urethral calibre.

Dr. F. J. Bumstead regarded internal urethrotomy as preferable to gradual dilatation in the treatment of stricture. Some time ago he was in the habit of treating urethral strictures by means of Holt's divulsor, but latterly and up to the time of his death, he discarded this method of treatment entirely. He found that internal urethrotomy was productive of better results when carried to a considerable extent than when more limited. In several cases he cut so as to admit a 35 or 40 mm. French scale. He thought that sounds larger than 25 mm. were constantly needed in practice. He thought that the meatus was slit up entirely too much, but he had never seen any bad results follow such slitting.

Dr. E. L. Keyes thinks that internal urethrotomy is better than dilatation in strictures situated in the anterior portion of the urethra. In the treatment by dilatation he prefers instruments below rather than above 30 mm. Strictures in the deeper portions of the urethra he treats by gradual dilatation and traumatic linear strictures by external section.

STRANGULATED AND INCARCERATED HERNIAS.

Dr. Frank Hastings Hamilton has reached the following conclusions regarding posture in the treatment of strangulated and incarcerated hernias:

1. Taxis is of prime importance.
2. Internal traction is only second to this in value. This is accomplished by securing the paralysis of the abdominal muscles and exciting peristalsis in the intestines.
3. Chloroform, hot baths and other similar agents are the best means for accomplishing muscular relaxation, peristalsis and anti-peristalsis.
4. Ice can only relieve the "button-holing" when this is due to congestion, and when it is applied very early. Opium is also of a somewhat limited application.
5. Emetics are of service in causing an upheaval of the viscera and in exciting peristalsis.
6. Purgatives act by causing peristalsis above, and anti-peristalsis below, the seat of stricture.
7. Stimulating enemata and enemata of tobacco also produce peristalsis, and are both direct and indirect in their effects.

8. All positions of the patient are beneficial in which the viscera are drawn unwards; and that is likely to be of the most service which causes the most efficient inward traction, at the same time that it does not interfere with the application of taxis.

PROLAPSE OF THE RECTUM.

In recent cases, Dr. Abram Jacobi finds an injection of ice-water sufficient to effect a cure. Where this fails, he resorts to zinc or alum. He finds the nitrate of silver, either in stick or solution of varying strength, an excellent remedy. In whichever form used, its action is immediately neutralized. Otherwise, it may give rise to extreme tenesmus. The actual cautery is employed with good effect in some cases. When there is only a paralysis of the sphincter ani, nux vomica or strychnia are employed. He very often employs strychnia hypodermically, using about $1/60$ of a grain every day. An ointment which he very often uses is one composed of ʒj of the alcoholic extract of nux vomica rubbed up with an ounce of fat.

LIGATION OF THE LINGUAL ARTERY IN THE REMOVAL OF CANCER OF THE TONGUE.

Dr. George F. Shrady reaches the following conclusions in this connection:

1. Cancer of the tongue, whenever possible, should be removed through the mouth.
2. Ligation of the lingual artery should always precede this operation.
3. This ligation should be performed near the origin of the vessel.
4. The use of the scissors and knife places the wound in a condition more favorable for rapid healing than when the écraseur or any variety of cautery is used.

5. Ligation of the lingual helps to prevent the return of the disease.

SCIRRHUS OF THE MAMMA.

Dr. C. K. Briddon operates by the electrolytic plan of treatment. A needle, attached to the negative pole of a Drescher constant battery, is plunged into the tumor near its base; a sponge electrode attached to the positive pole is applied to the opposite side of the tumor, and the current from ten cells set in action. No anæsthetic is used.

EXTIRPATION OF THE RECTUM FOR CANCER.

This operation is performed as follows, by Dr. Keyes: The mucous membrane is incised from the tumor downwards, and the incision carried, externally, through the skin only as far as the tip of the coccyx. A stout needle and ligature, carrying the wire of the écraseur, are then passed from near the tip of the coccyx, upwards, outside of the gut, above the disease, then through the wall of the bowel and out at the anus. The tissues included in the loop of the wire are then slowly divided. The diseased portion of the gut is now drawn well down, and stout knitting-needles are thrust through the healthy tissue, beyond it, on each side, successively. The loop of the écraseur is then carried round, above the needles, and the whole mass thus removed. No antiseptics are used.

FIBROUS ANCHYLOSIS.

Inflammatory action is subdued, by Dr. Lewis A. Sayre, by rest in bed and continued application of cold to the inflamed parts by means of ice-bags. When inflammatory action is subdued, the patient is placed upon an instrument consisting of a pelvic belt, with perineal bands; a long bar, with a foot-piece and adjustment, for extension; a knee-cap, and a movable joint, opposite the hip, for flexion, extension and abduction. The movable joint is so arranged that abduction and rotation of the limb outwards can be effected at the same time. The most admirable effects have, in most cases, followed the use of this apparatus.

CHRONIC SYNOVITIS.

Decided benefit has been obtained, in the sub-acute stage, by pressure applied to the joints. This is accomplished by means of a compressed sponge. The joint, which is the seat of the disease, is covered with a compressed sponge, which is retained in position by means of a roller bandage. The sponge is then wet with warm water, which causes it to gradually expand, and thus produce an equal amount of pressure over all the parts covered. The sponge is applied once or twice a day, as the case may demand.

To remove the fluid, simple aspiration, followed by elastic pressure, suffices in most cases, but where simple aspiration does not succeed, Dr. Sayre removes the fluid by means of the ordinary trocar, and then injects the cavity with Lugol's solution of iodine. After this operation, the patient is placed in bed, the knees firmly bandaged, locked in a perfectly immovable apparatus, and elevated above the rest of the body. Ice-bags are employed, if necessary, so as to keep down inflammatory action.

This disease is treated at the Roosevelt Hospital by leeches applied over the diseased joint.

CANCER OF THE LOWER LIP.

Dr. Sayre removes these growths at once, with a knife. He avoids hemorrhage by having an assistant make pressure upon the facial arteries, as they pass over the ramus of the lower jaw. The wound is closed by means of sutures, or pins with the figure-of-eight suture. Pins are most frequently used, and are passed through the lips of the wound, whose edges are then brought together in such a manner as to avoid leaving any notch in the free margin of the lip. The attachments of the cheek are loosened with the knife, if necessary, in order to give more opportunity for perfect adjustment. Two pins have usually sufficed. After the pins are adjusted, and the sutures twisted about them, a piece of adhesive plaster is placed beneath the point of the pins, to prevent irritation and excoriation. Then long, narrow strips of adhesive plaster are adjusted so as to give support to the pins in holding the lips of the wound in co-aptation. These strips, passing above and below each pin, are carried far back upon the sides of the face and neck. The pins are removed within 48 hours, at most, after the operation. They are removed by seizing them at the head with a pair of pincers and carefully turning them round once or twice, before making the least traction. In this way, the pins are withdrawn without disturbing the threads or plasters, which, together with the crusts, are left remaining, and are not removed for some time.

INGROWING TOE-NAILS.

Immense relief has been afforded in these cases, by Dr. Sayre, by applying a few threads of cotton beneath the cutting edge of the nail, in such a manner as to protect the excessively-tender tissues from the irritation produced by being brought into close contact with it. This cotton is applied by means of a narrow, thin-bladed knife, without cutting edge. With this instrument, he draws a few threads of cotton down between the nail and the mass of granulations, and so on until they are carried beneath the cutting edge of the nail. After placing the threads *in situ*, the fungous granulations are penciled over freely with nitrate of silver. This application is repeated as often as the destroyed tissues separate, until the exuberant growth is all destroyed.

INFLAMMATION OF THE WRIST-JOINT.

The indications consisting in placing the joint perfectly at rest, and at the same time removing all pressure from the articular surfaces, Dr. Sayre proceeds as follows: He takes a piece of sole-leather, long enough to half or two-thirds surround the arm. This is then dipped in cold water, and made thoroughly flexible. It is then covered with a piece of adhesive plaster, plaster side out, long enough to go completely round it, lengthwise. Each opening is then covered with a piece of oakum. The leather-lined plaster he now applies to the palm of the hand, moulding it and securing it with a roller bandage, as far as the wrist. Grasping the hand already covered, while an assistant grasps the arm near the elbow, he makes extension and counter-extension, until the patient says that all pain is relieved, and then brings the remainder of the leather-lined plaster against the forearm, and secures it with a continuation of the bandage. In this manner, all pressure is removed from the articular surfaces, pain is relieved, and an apparatus is afforded which retains everything at perfect rest.

FISTULA IN ANO.

Several cases of this disease have been treated in the wards of the Roosevelt Hospital by means of the elastic ligature, and the patients have suffered but little inconvenience while the ligature was cutting through. They were up and about most of the time, and suffered but little pain. In one case the ligature cut its way through in five days; in two other cases, in seven days, and in every case a fine granulating surface was left behind, which healed much more readily and satisfactorily than the wound made by the knife.

EMPYEMA.

The treatment pursued at the Roosevelt Hospital, in cases

of this disease, is by free incision through the chest wall. The pleural cavity is then washed out with a solution of salicylic acid, of the strength of 1 part to 500 of water. If the opening shows a tendency to close at any time, it is dilated with sponge-tents.

BORACIC ACID AS AN ANODYNE.

Pain in and about the wound, subsequent to removal of the breast for scirrhus, is allayed, at St. Luke's Hospital, by applying cloths which have been wet in a solution of boracic acid.

SILICATE OF LIME SPLINT.

This splint is obtained by using a saturated solution of chloride of lime and the silicate of soda. It is supposed that a chemical decomposition occurs between the two articles, which gives rise to silicate of lime. The apparatus is made in the following manner: First apply a bandage which has been soaked in the solution of lime and then squeezed very nearly dry; next give the bandage applied a coating of silicate of soda; then apply another bandage wet in the lime solution and follow it with a coating of the soda. Thus go on until the requisite number of bandages are applied. The splint, when dry, is much harder than the silicate of soda splint. It is much lighter than a plaster splint, and has given general satisfaction at Bellevue Hospital.

SURGICAL DRESSING.

The dressing has been found, at Bellevue Hospital, to be a valuable one for all granulating surfaces.

R Basilicon ointment... ℨj.
  Balsam of Peru .... ℨj.   M.
  Sig.—Apply spread upon lint.

BURNS.

At Bellevue Hospital burns are coated with mucilage of gum arabic and are then dusted with lycopodium. As soon as any of this dressing falls off it is re-applied. In other instances, a solution of the nitrate of silver (gr. xx to the ounce) is employed. In some cases, solutions as strong as gr. xl to the ounce have been employed with benefit. The solution is applied with a camel's-hair brush, and the surface is left exposed to the influence of light and air. The solution is renewed whenever fissures occur in the coating thus formed.

ANTHRAX.

A very simple and efficient method of management has been adopted at Bellevue Hospital. A broad piece of spongio-piline is chosen and a hole cut through it large enough to receive the apex of the tumor. It is then applied, and the anthrax let alone.

ABSCESSES

At Bellevue Hospital are at first injected with a solution consisting of equal parts of water and tincture of iodine. This solution is gradually increased in strength until pure tincture of iodine is employed.

INTERNAL HÆMORRHOIDS.

Dr. Erskine Mason divides treatment into medical and surgical. The aperients which he considers most valuable are the salines in combination with sulphur, or a pill composed of taraxacum and aloës.

As regards the employment of nitric acid, he thinks that its use should be confined solely to those vascular spots of mucous membrane which are sometimes seen in connection with other tumors, and to the small, florid, sessile growths, which so readily bleed upon the slightest touch. The acid which he employs in such cases is of the strongest kind. The parts are dried and then touched lightly with a piece of wood dipped in the acid, avoiding the surrounding mucous membrane. The parts are then oiled and returned. One or two applications generally suffice.

In some cases Dr. Mason employs the clamp and actual cautery. He grasps the pile with forceps, or tenaculum, drags it down, and strongly compresses it by means of a clamp around its base. The pressure is maintained by means of a screw. He then, with a pair of curved scissors, clips off the pile, a short distance from the clamp, so as to leave a stump, over which an iron heated to a dull heat is drawn.

He regards the ligature as the safest and most efficient means of treating piles. He employs a moderately fine, waxed, silk ligature, or one of linen. After seizing the tumor and dragging it down, he surrounds it with the ligature, ties it tightly in two knots, and cuts it off a little distance from the knot. He is careful *not* to tie the ligature *close up* to the base of the tumor, and so avoids including the coats of the intestine and consequent troublesome contraction of the bowels.

He regards with great favor Mr. Allingham's method of treating internal hæmorrhoids which consists in separating the pile, with the scissors, from its attachment to the muscular and other tissues of the bowel beneath its mucous membrane. The cut is carried up for a little distance parallel to the wall of the bowel and the neck of the tumor is then ligated. In this way little more than the vessels which form it are tied · and there being less tissue for the ligature to separate, it comes away sooner. The vessels running parallel to the incision are not likely to be wounded, and if there is any bleeding point it is readily seen and can be tied at once. The wound, being an incised one, readily heals. After this operation the patient is confined in bed for at least a week. Dr. Mason is in the habit not only of having the anus frequently bathed with warm water, but also irrigates the parts with tepid water and a little carbolic acid by means of a small syringe. Immediately after the operation a pad of picked lint is placed over the anus and held in position by a tight T bandage.

WET STRAPPING.

It is regarded at Bellevue Hospital as an item of some importance in the treatment of old ulcers, such as are of specific nature, found upon the lower extremities, that, if strapped, the plaster should be permitted to remain as long as possible without change. With this object in view several cases have been dressed with what have been called wet straps, which are prepared by passing strips of adhesive plaster through hot water instead of heating them in the usual manner. In these instances the water is also carbolized.

It seems quite certain that the ulcers heal more rapidly under this plan of treatment than any which has been adopted, and that the plaster does not get loose as quick as when heated over a spirit lamp.

ECZEMA RUBRA.

A number of cases of this disease as occurring in the wards of Bellevue Hospital have followed facial erysipelas, and, after trying all the remedies ordinarily resorted to in the treatment of this affection without benefiting the patient, tincture iodine was painted over the entire surface. As a result, the disease was in every way aggravated for one or two days, but when the artificial irritation subsided, the change produced in tissues by the iodine permitted the cases to go on to rapid and complete recovery.

DOUBLE COLLES' FRACTURES

Are dressed at Bellevue Hospital with straight board splints

6 or 8 inches long, the anterior splint reaching down to the joint, the posterior 1, 2 or 3 inches upon the dorsum of the hand. These splints are then secured with roller bandage. Passive motion is begun at the end of the third week.

THE TREATMENT OF FRACTURES AT THE NEW YORK HOSPITAL.

Colles' Fracture.—The Morris treatment is that generally pursued. Two bandages, 3 inches in width and rather loosely rolled, are placed on either surface of the forearm, over the seat of fracture, and maintained in place by a broad strip of adhesive plaster, making sufficient compression to keep the fragments in position.

Fracture of the Patella.—Tilford's strap-and-buckle apparatus is applied. After union has taken place an elastic knee-cap is used.

Fractures of the Tibia and Fibula.—The fracture-box is first employed and followed by the plaster-of-paris bandage.

Fractures of the Femur are treated either with the plaster-of-paris splint, or with Buck's extension apparatus, with Volkman's sliding contrivance for keeping the foot straight, and ordinarily a long external splint.

EPIDIDYMITIS.

When a case of gonorrhœa presents itself at Bellevue Hospital the patient is made to wear a suspensory bandage. This bandage may prevent swelling of the testicle. When the epididymitis occurs in spite of the bandage and is seen early, if the pain is severe, needle punctures are made in the tunica vaginalis, and the hydrocele, which causes the pain, thus relieved. As a rule it has not been found advisable to use urethral injections during the acute stage of epididymitis. In this stage sedative lotions, fomentations, leeches, and cold applications are employed. Rest in bed is insisted upon. If leeches are used they are applied along the course of the spermatic cord. The testicles are supported by passing a broad strip of adhesive plaster under the scrotum and from one thigh to the other.

THE TRANSPLANTATION OF SKIN.

Dr. H. D. Noyes lays great stress upon the observance of the following rules:

First. The piece of skin used must be dissected so as to be entirely free from subcutaneous connective tissue and fat.

Second. The piece of skin should not be secured in position by means of sutures, but by co-aptation with gold-beater's skin, or such means as maintain absolute immobility during the entire process of union.

Third. There must be expected an extraordinary contraction when the piece of skin is separated from its surroundings.

ACUTE, NON-SPECIFIC URETHRITIS.

Rest, and 30 grains of the citrate of potassium three times a day, is the treatment generally pursued with success by Dr. Gouley.

PERINEAL FISTULA.

Firm pressure is made with the finger, placed either in the rectum or at the point of the internal orifice of the fistula, every time that the patient urinates, in order that the urine may pass through the urethra, and none escape into the fistulous tract. In this way, Dr. Gouley gives the fistula an opportunity of healing up by granulation, without having the process interfered with by the passage of urine over the surface.

CHANCRE.

When a well-determined initial lesion is situated in loose

tissue, Dr. F. N. Otis excises it as early as possible. The infective neoplasm he removes *entire*. He performs the operation of incision in the following manner: The parts are first thoroughly cleansed by gentle bathing in warm water. In all open lesions, a solution of carbolic acid, of the strength of 1 part of the acid to 40 parts of water, is applied, after which, the indurated mass is raised between the forefinger and thumb, and encircled firmly at the base with a bit of fine silver or malleable iron wire. Then, with a narrow, sharp-pointed bistoury, the tissues are pierced at the centre, beneath the compressing wire, or probe, and cut well under and out, including all the indurated and a little of the sound tissue of that side. The same cut-out is then made on the opposite side. After every portion of the neoplasm is removed, interrupted sutures of silk, or silver wire, are then introduced at intervals of one-fourth of an inch. The patient is kept in the recumbent position, the parts being occasionally wet with carbolated water, until the third day, when union by first intention, is usually found to have taken place.

When excision is unadvisable by reason of the location of the indurated papule, it is subjected to applications of the oleate of mercury (six per cent. solution), or a mild mercurial ointment, is employed. When the mucous membrane covering the induration is abraded, it is dusted with dry calomel and protected with a thin layer of dry lint. Calomel, in combination with lime-water, in the proportion of a drachm to a pint, or half a drachm of the bichloride of mercury, are much esteemed by Dr. Otis as topical applications. Calomel, in some form or other, is always administered internally.

The *mucoid Chancre* is treated by the application of the solid nitrate of silver.

The *inflamed, or suppurative initial lesion* is treated by rest and some sedative dressing such as the lotio plumbi et opii, in a 5-grain solution, or the simple powdered iodoform, or iodoform with an equal part of tannic acid.

The *gangrenous form* is managed by poultices of powdered

charcoal, and by the internal administration of mercury Where the patient is debilitated or scrofulous, generous diet, quinia and iron are necessary.

Concealed initial lesions are treated with bougies or suppositories, medicated with opium, salicylic acid, or iodoform.

THE CHANCROIDAL ULCER.

Here are some formulæ which Dr. Sturgis recommends for local application:

1. ℞ Pulv. iodoformi... 1 part.
      Lycopodii.. 2 parts. M.
   Triturate well, apply locally.

2. ℞ Pulv. iodoformi,
      Pulv. acid tan.. p. œ. M.
   Triturate and use locally.

3. ℞ Pulv. iodoformi... ʒj.
      Tinci sulphat.. gr. v.
      Pulv. acid. tan... ʒj. M.
   Triturate. For local use.

4. ℞ Acid. carbol. cryst....................................... ʒj.
      Aquæ.. f℥viij. M.

Or,

5. ℞ Zinci sulphat.. gr. v-xx.
      Aquæ.. f℥ ij. M.

6. ℞ Acid. nitrici... fʒ ss.
      Aquæ.. fʒ viij. M.

As agents destructive of the chancroidal ulcer Dr. Sturgis employs (1) the *white iron,* or the *galvano cautery,* (2) *strong sulphuric acid,* (3) *chemically-pure nitric acid,* (4) *pure carbolic acid.* He very generally applies sulphuric acid in the shape of Ricord's carbo-sulphuric paste, which is made by taking a small quanity of finely-powdered *willow* charcoal, and enough of the acid to make a paste of the consistence of thick cream. This is put on with a porcelain or glass spatula, care being taken to carry the agent into sound tissue both beneath and on the surface of the edges of the chancroid.

Salicylic acid is first sprinkled over the sore at the New York Hospital, and then the following wash applied:

℞ Acid salicyl.,
Sodii borat.. āā gr. xv.
Glycerinæ.. f℥j.　　M.

THE TREATMENT OF HÆMORRHOIDS.

For the purpose of effecting a radical cure, piles are treated at Bellevue Hospital with the double ligature. The sphincter is not dilated. Treatment after the operation is believed to be sufficient to prevent spasm of the sphincter. In cases in which the pile can be easily surrounded, Sims' speculum is recommended. The operation being completed, a suppository containing opium and belladonna is introduced well above the internal sphincter. A large piece of lint smeared with vaseline is then placed over the anus and the cleft of the nates, and the cleft packed with cotton until it is filled to a level with the tuber ischii; over that a compress is placed, and the whole is retained in position by means of an ordinary T bandage.

But if *hemorrhage* occurs from the stump of the pile or elsewhere, a plan of treatment is recommended, which although not new, is perhaps worthy of description. Of course, effort is made to tie the bleeding vessel. But the plan recommended as the better one is first to take a cone-shaped piece of sponge, and make it hollow; then to pass a thread from the inside through the side of the sponge, over the apex of the cone, and return it to the cavity in the sponge. In that manner a loop is made which places the sponge within the control of the surgeon. The sponge is to be slightly moistened, compressed, and pushed up as high as possible in the rectum upon the tip of the finger. Pieces of lint are then carried in until the cavity in the sponge is filled. As soon as filled, traction is made upon the strings, when the sponge will spread out and press against the sides of the rectum. In this manner the flow of blood upward is prevented, and the compress already described prevents any discharge from the anus. In ordinary cases it is thought advisable to leave the sponge *in situ* for 36 or 48 hours. If hemorrhage returns, the sponge is replaced.

ACUTE RETENTION OF URINE.

A safe rule, therefore, for guidance in the management of cases of acute retention of urine of forty-eight hours' duration, which Dr. Gouley lays down is, never to draw off more than one-third of the contents of the bladder, and to do this very slowly by half closing the distal end of the catheter, so that the urine will flow in a very small stream. Having collected half a pint, he closes the catheter for a quarter of an hour, then lets another half pint flow, and so on, until the required quantity has been obtained. In two hours he repeats the catheterism, if the first has been easy—otherwise, the catheter is closed, and left in for twenty-four hours—and removes again the same quantity very gradually, and at the expiration of another period of two hours completely empties the bladder, always slowly; and in this way he takes the necessary precautions to avoid both cystorrhagia and polyuria. Every three hours after the last catheterism the urine is drawn off until the patient can pass it spontaneously; if he cannot do so, the catheter is resorted to at such intervals as are found necessary. Dr. Gouley also finds it necessary to treat the existing vesical inflammation and atony. For general medication he recommends the tincture of chloride of iron, in 5-minim doses, three times daily, and also diluent drinks, such as 30 grains of citrate of sodium or potassium, in half a glass of water three times daily, and in a few days, for a change, doggrass tea, etc. Topically, lumps of ice the size of the last joint of the thumb, are introduced into the rectum in rapid succession as fast as they melt, for an hour, night and morning. A bag of ice is afterwards applied alternately to the perinæum and hypogastrium for an hour or more. Each time that the bladder is emptied by the catheter, a couple of ounces of cold borax solution, from 5 to 10 grains to the ounce, are thrown in and allowed to run out slowly, then two more, and two more ounces which are left in and the catheter withdrawn; this is accomplished in five minutes. Dr. Gouley has found that no preparation gives more satisfaction than the borax solution for cleansing bladders which contain offensive purulent urine. He thinks it is as well to have in readiness a strong solution of borax in glycerine; 1 ounce of the biborate of sodium being readily dissolved by 6 ounces of glycerine, each drachm of such a solution being equal to 10 grains of biborate. In a week, or thereabouts, if the case progresses well, the irrigations are diminished in number until only one is used each day. In some cases a mild faradic current has been found serviceable.

CYSTITIS FOLLOWING GONORRHŒA.

Dr. Gouley regards this as a very troublesome affection, and one which is apt to last from three to six months. Some of the French surgeons, he says, are in the habit of treating it by injecting 5 or 10 minims of a solution of nitrate of silver, (30 or 40 grains to the ounce,) from time to time, into the bladder; but personally, he prefers the internal treatment by means of diluents and balsamics. In addition, he thinks it well to give belladonna, if the drug is well borne, and begins with about $\frac{1}{4}$ of a grain three times a day. In some cases, even smaller doses than that have, in his hands, sent the pulse up to 120, and produced the most uncomfortable symptoms, while other patients can take from 2 to 5 grains a day with impunity. In some instances he is in the habit of combining the aqueous extract of opium with the belladonna, (a grain of the one to ¼ of a grain of the other,) and sometimes he finds that rectal suppositories containing these ingredients act very admirably.

CYSTORRHAGIA.

As a preventive treatment of cystorrhagia, and in cases of over-distention of the bladder, Dr. J. W. S. Gouley teaches that the bladder should be emptied very gradually in the course of from 12 to 72 hours. This is accomplished by the introduction of a soft catheter, which is secured in position, and the urine then allowed to trickle through it slowly, by partially obstructing the distal orifice with a finger, so that, in the course of five or six minutes, not more than eight ounces will have escaped. This process is repeated every two hours, until the bladder is empty. If the urine drawn is very fœtid, the process is modified as follows: Immediately after drawing the first half pint of urine, eight ounces of warm water, in which has been dissolved a scruple of biborate of soda, are injected into the bladder, then a pint of urine drawn, and immediately, half a pint of borax solution thrown in. This procedure is repeated six or eight times, at the first sitting, or until the whole of the offensive urine is gotten rid of, and the bladder contains nothing but the clear borax solution; then, every two hours, and in some cases, even every three hours, half a pint of the fluid—now mixed with newly-secreted urine—is drawn off, until the bladder is completely empty. In this way, a bland fluid is substituted for decomposed and irritating urine, the hydrostatic pressure will not have been removed too suddenly, and cystorrhagia is prevented.

When cystorrhagia has taken place, besides enjoining absolute rest, in the horizontal posture, Dr. Gouley takes means to prevent the bladder from becoming distended with bloody urine. If he finds hypogastric dullness extending to or into the umbilical region, he draws off, through a gum catheter, only a pint of urine, then throws in half a pint of the borax solution, and draws half a pint of fluid, and repeats the process until the fluid becomes clear, or nearly so. Afterwards, every hour or two hours, the stopper of the catheter is removed, and half a pint of fluid allowed to flow, until the bladder is completely empty. The catheter is retained in position for 24 hours. The hemorrhage is checked by occasionally throwing into the bladder two or three ounces of a weak solution of tannin, or of alum. Dr. Gouley has also found ice in the rectum a valuable adjuvant.

He treats hemorrhages caused by papillomatous or cancerous tumors, by the exhibition of ergot, in the shape of from ♏xv-xx of the fluid extract, given every two or three hours, or else he uses gallic acid dissolved in glycerine, or quinia with an equal quantity of dilute sulphuric acid, or tincture of the chloride of iron, etc. The dose is reduced and altogether discontinued, as soon as possible, so as to save the digestive organs.

When no fluid runs through the catheter, after it has been properly introduced, owing to the fact that a clot has become impacted in the catheter, Dr. Gouley removes the catheter and washes it out, or takes a clean catheter and introduces it. But when the bladder is filled with clotted blood, he uses a soft catheter, with one large eye, or a metallic Mercier catheter, as large as can be introduced, and injects one or two ounces of warm borax solution. If nothing flows, the instrument is moved gently to and fro, and rotated to the right and left, so as to break up the clots. Aspiration is then made, either with a syringe or with Bigelow's rubber bag. Not more than three or four ounces of clots, or grumous clots, are aspirated at a time. Alternate injections and aspirations are made at the same sitting.

THE MEDICAL AND SURGICAL DISEASES OF WOMEN.

VAGINISMUS.

Dr. T. G. Thomas treats this condition as follows: First, the patient is thoroughly anæsthetized with ether, and then each *labium majus* is held back by an assistant. The parts being thus exposed, the hymen is seized with a pair of mouse-toothed forceps, and snipped completely out with the scissors, which are preferable to the knife for this purpose. The sponge is usually all that is necessary to stop the bleeding. The entrance of the vagina is then cut down upon and enlarged, and while this is being done the assistants are instructed to stretch the vagina well on either side. The incisions are made into the perinæal body, but not through any muscle. Three incisions are all that Dr. Thomas finds necessary. Of these, one is in the median line, and one on either side. One of Sims' plugs is then pushed into the vagina and held in place by means of a broad strip of adhesive plaster passing from the lower part of the back, over the perinæum and up to the abdomen, with a hole cut in it for passing a catheter. The plug puts an end to all hemorrhage, and is generally left in position for three or four days before being disturbed. At the end of that time, it is taken out so as to permit the vagina to be thoroughly syringed with warm water. It is then replaced as before. In a week's time the patient is able to remove it and put it back herself. It is to be worn for three weeks or more constantly, then it is only necessary to wear it at night. In the course of a month or six weeks, Dr. Thomas' patients are able to dispense with it altogether.

LABOR IN KYPHOTIC PELVES.

Dr. Isaac E. Taylor reaches the following conclusions:

1. A mutilated child can be delivered with safety to the mother through a space of 1¾ inches antero-posterior, and 2½ to 3 inches transverse diameter, by craniotomy, cephalotripsy, or cranioclasm. When the vault has been destroyed, the face is made to present edgewise, or the head sidewise.

2. After cephalotripsy, or cranioclasm, if necessary, version with propulsion from above the pubes, performed early and before the uterine forces are exhausted, is preferable to that just indicated.

3. The cephalotribe, or cranioclast, cannot be considered as available tractors in cases of extreme contraction of the pelvis, but other instruments become necessary to properly effect the delivery of the woman.

3. The Cæsarean section should not be performed when contraction or deformity is present, as stated above, unless demanded by other conditions and complications.

LACERATION OF THE PERINÆUM.

Dr. Montrose A. Pallen urges immediate operative interference, if possible. He enters his sutures (silver wire) deep, and not more than a line apart, and supplements them by superficial sutures in each inter-space. The bowels are kept soluble by warm enemata, and not constipated by opium, after the common rule. The vagina is kept thoroughly cleansed, the bladder well emptied, and the nates of the patient are placed in an india-rubber ring cushion filled with air.

MENORRHAGIA.

Dr. Barker directs his patients to use for a week previous to the expected period, the following suppositories:

℞ Ext. ergot. aq. (Squibb)............................... ℈ij.
   Cacao butter.. ʒj.   M.
   Et in suppos. No. xij div.
   SIG.—One to be introduced into the rectum morning, noon and night.

These suppositories are carried as far as possible up into the bowel, and the patient directed to remain in a recumbent position for at least an hour after using them.

In cases of irregular uterine hemorrhage seen in connection with the climacteric period he introduces into the cavity of the uterus cylinders of iodoform made according to this formula:

℞ Iodoform .... ʒijss.
   Gum tragacanth.. gr. xv.
   Mucil.. q. s.   M.
   Et div. into cylinders No. x, each one and one-half inches in length.

One of these cylinders is directed to be carried completely into the cavity of the uterus, and a pledget of cotton is placed against the cervix to retain it in position. One of these cylinders is introduced daily for five or six days previous to menstruation.

POST-PARTUM HEMORRHAGE.

When this is due to *inertia of the uterus,* Dr. Montrose A. Pallen introduces the hand into the cavity and removes the clot, then incites contractions of the organ by manual pressure, and administers stimulants and ergot hypodermically.

Where consequent to *laceration of the cervix uteri*, the tampon is used and stitches are necessary.

PRURITUS VULVÆ DUE TO VAGINAL LEUCORRHŒA.

Dr. Thomas recommends frequent vaginal injections of the biborate of sodium in solution, and once or twice a week he cleanses the cervix thoroughly of mucus and applies the nitrate of silver. Occasionally he uses chemically-pure nitric acid with the hope of altering the secretion. Copious injections of water are continually used and the patient is told to press a suppository of butter of cacao containing 5 grains of tannic or gallic acid up against the cervix twice daily.

ABDOMINAL PREGNANCY.

Dr. T. G. Thomas lays down the following rules for the management of cases of abdominal pregnancy:

1. Before full term, if the child is alive, its growth may be carefully watched with the hope of delivering a living child at the end of the ninth month by the operation of leparotomy and also of saving the life of the mother.
2. If the child dies early in abdominal pregnancy, delay is advisable, but it should not be carried to the extent of the development of hectic and septicæmia.
3. At full term the best rule is to await the evidence of constitutional disturbance, and then meet its development promptly by operative interference.

Dr. Fordyce Barker lays great stress (1) upon leaving the placenta *in situ* after the operation, and (2) upon subsequent antiseptic treatment.

LACERATION OF THE CERVIX UTERI.

Dr. Gillette has recently resorted to a somewhat novel method of treating this condition. He seizes the torn cervix with two tenacula, draws the lacerated surfaces together, and then slips over the handles of the tenacula and around the cervix an ordinary rubber strap. This treatment is not applicable of course, to recent lacerations, since in such cases the band constringing the cervix would prevent the discharge of the lochia.

SUBMUCOUS AND INTERSTITIAL FIBROIDS OF THE WOMB.

Dr. Thomas believes that enucleation and the use of ergot

are attended by the great dangers of septicæmia, peritonitis, hemorrhage and exhaustion. He is, on the other hand, convinced that a policy of inactivity is by no means always a safe one. He thinks that the various other methods of treatment employed at the present day, *excision, torsion, avulsion, ecrasement, and the production of sloughing*—are none of them without objection, and offers in their place the following method, which consists in seizing the most dependent and accessible part of the tumor with a strong Volsella forceps, passing along its sides the *serrated scoop* or *spoon-saw*, and by a gentle pendulum motion from side to side, sawing through the attachments of the tumor and forcing it entirely from its connection with the uterus.

He claims the following advantages for the instrument:

1. The attachments are separated by a saw which greatly limits hemorrhage.

2. The shape of the spoon, convex without and concave within, causes it to follow, of its own accord, the contour of the tumor and at the same time to protect the uterine tissue.

3. The highest attachment can be as readily reached as the lowest.

4. The saw action secures separation with rapidity and with certainty.

5. The spoon-saw secures separation of the growth at its highest point of attachment, and leaves no particle to decompose.

In order to determine the extent of the attachment of a tumor, Dr. Thomas uses a *flat whalebone* sound. He uses this instrument in this way: the index finger of the left hand is placed against the most accessible part of the tumor, and then the sound is passed up along the side of the tumor until it is arrested. The sound being then withdrawn with the finger still upon it, is laid upon a sheet of paper, and being curved, a line is drawn from its tip to the point touched by the finger. The same is done for the opposite side of the tumor.

In the delivery of large tumors from the vagina, after their expulsion from the uterus, he recommends the following procedures:

1. Seize the tumor with a strong forceps, draw it down, sever the distended perinæum to the sphincter ani, partially or completely invert the uterus, detach the tumor by means of the spoon-saw, replace the uterus at once, and close the perinæum by sutures.

2. Successive sections of the tumor may be cut away by means of the galvano-caustic wire.

3. A large trocar and canula, or the actual cautery, or the trephine obstetric perforator, may be used to channel up the middle of the tumor, and then, with a strong pair of scissors, or with the osteotome, pieces can be cut out, and the tumor so diminished in size that it is susceptible of delivery.

Dr. Alfred Post finds much advantage arising from passing a strong ligature through the part of a large tumor which projects from the vagina. This is a more powerful means of making traction than by the use of the Volsella forceps alone.

In the removal of large tumors which have been driven into the vagina, Dr. Thomas Addis Emmet does not divide the perinæum or enter the uterine canal, at all. He believes that the fact that such tumors are in the vagina, shows that the uterus is strong enough to drive them completely out.

ABORTION.

Dr. W. T. Lusk thus summarizes his views on this subject:

1. In the first two months, an abortion needs no special treatment. The hemorrhages of early date are amenable to the same principles of treatment as those from the non-pregnant uterus.

2. In the third month, no treatment is required, when the ovum is expelled with intact membranes. When the membranes rupture previous to expulsion, and hemorrhage takes

place, immediate removal should be attempted, provided that the cervix be sufficiently dilated to admit the index finger. When the cervix is closed, the tampon should be tried for 24 hours. If the tampon proves ineffective, the cervix should then be dilated with a sponge-tent, and the ovum removed with the finger. The finger should be made to pass up along the side of the uterus, across the fundus, and so complete the circuit of the uterine cavity.

3. In cases of neglected abortion, retained portions should be removed by the finger or the curette. When the ovum is decomposed, no dilation of the os is usually needed. When the ovum is fresh, the preliminary use of sponge-tents is usually demanded, if manual delivery is resorted to.

4. Fibrinous polypi, when situated near the os internum, arrest the involution of the lower portion of the uterus. The os is therefore open, and permits the passage of the finger. When the polypus is attached to the fundus, the cervix is usually closed. Small, smooth, slippery bodies, like fibrinous polypi, are rarely to be detached, unless the finger operates from above so that the choice of hands depends upon the side to which the polypus is attached.

5. In immature deliveries, hemorrhage can usually be controlled without the tampon, by compression of the uterus, and, in cases of delay, by the manual extraction of the placenta.

Dr. Fordyce Barker believes that, occasionally, treatment is required in abortion occurring in the first two months of pregnancy. In these early abortions, he is accustomed to inject into the vagina a very large quantity of very hot water, from 104°–110° F. He is sure that this will absolutely arrest the hemorrhage.

In cases of abortion where it is necsssary to tampon, he does not trust to any kind of vaginal tampon, but always plugs up the cervix uteri with a compressed sponge-tent, and then only fills up the vagina sufficiently to keep this sponge in place.

SORE NIPPLES.

Dr. F. V. White is in the habit of simply protecting the nipple with an ordinary nipple-glass, secured in position by means of a bandage.

Dr. S. S. Purple uses the following:

R Tannin.. ʒj.
    Syr. acaciæ.. f℥ ij.
    Aquæ... f℥ ij.   M.

This is applied to the nipple and breast with the finger, and allowed to remain exposed to the air until perfectly dry.

Dr. Compton uses this formula:

R Tr. benzoin. co.,
    Glycerinæ..āā q. s.   M.

OVARIOTOMY.

If the operation can be safely delayed for a week or more, after coming under treatment, Dr. Nathan Bozeman prepares the patient by administering to her tonics and food as much as she can bear. Iron he considers a most valuable agent in the preparatory stage of the treatment. The antiseptic method (Lister's) he invariably uses in this, as in all major operations. He thinks his successes are greatly due to the means thus adopted of preventing peritonitis and septicæmia. Whether long or short, he returns the pedicle into the peritoneal cavity, after transfixing and tying it, right and left, several times with waxed, carbolized, strong silk ligatures, and claims that there is no necessity of using clamps or Koeberle's serre-næud. He includes the peritoneum in his sutures when closing the abdominal *incision*, which he never makes larger than is necessary in the median line. Carbolized silk sutures are also used for closing the wound as for tying the pedicle. Beef tea, milk and eggs constitute the food given as soon as the patient has fully recovered from the anæsthetic, (ether being used for this purpose.) If there is a tendency to vomiting, the food is administered per rectum. Quinine and opium the doctor considers of the highest importance in the after-treatment, given in full doses, as being antiperiodic, and a preventive of peritonitis. Should there be an undue elevation of temperature, not controlled by the medication enumerated, Kibbe's cot comes into requisition. The first incision he never makes larger than is necessary for the introduction into the peritoneal cavity of his abdominal spatula, as the doctor terms it, (a flexible, metallic rod, 10 to 12 inches long, well rounded off, with a triangular-shaped termination at either end, like Nott's vaginal depressor,) about one inch long, also well rounded off. The size of the tumor, its adhesions, if there be any, are thus explored with the aid of this spatula. The incision is then enlarged to 4–6″, for the purpose of introducing the hand and separating the adhesions, if their presence has been made out, in the mode above described. The next step consists in tapping the cyst or cysts with Spencer Wells' trocar. In multilocular cysts he taps one cyst after the other, through the opening made in the first cyst, and so on, the patient being turned on her side. The cysts are thus emptied to a size sufficient to pass his right hand through the abdominal opening into the peritoneal cavity while drawing out the cyst or cysts with his left. This simultaneous use of both hands Dr. Bozeman considers of the utmost importance while drawing out the cyst. The right hand introduced inside the cavity completes the separation of adhesions that may have remained after the use of his spatula, and also guards against any undue stretching or possible rupture of the intestines, gall-bladder, etc., with which there may be adhesions. The omission of this precautionary measure doubtless has caused many fatal results that might have terminated favorably had this precaution been practiced. Six to 8 grains of quiniæ sulphate, and 25 drops of the liquor opii comp., administered per rectum, are the doses of these remedies used from the first for the purposes mentioned. The use of hypodermic injections is

avoided by Dr. Bozeman. After ovariotomy, he is of the
opinion that on account of the pain thereby produced, the
patients abhor them, and thus cause undue nervous excitement. Dr. Bozeman never uses drainage tubes through
Douglas' cul-de-sac, but prefers to draw off effusions by means
of tubes introduced through the abdominal opening, reaching
down to Douglas' cul-de-sac.

To control the high temperature occurring after ovariotomy,
Dr. T. G. Thomas recommends the use of "Kibbe's fever-
cot." This cot consists of a strong, elastic, cotton netting,
manufactured for the purpose, through which water readily
passes to the rubber cloth below, which is so adjusted as to
direct the stream into a vessel at the foot of the bed. His
method of managing his patients is as follows: He places
upon the cot a blanket or sheet, which is kept constantly wet
by pouring cold water upon it. The temperature is taken
every hour; bottles of warm water are applied to the feet and
hands; the patient is allowed to remain upon the cot as long
as necessary, constantly enveloped in the wet sheet, and constantly exposed to the influence of cold water. The urine is
drawn with a catheter, and a bed-pan is used. His aim is to
keep the temperature of the body at 100° F., or a little less.

PUERPERAL CONVULSIONS.

Dr. S. T. Hubbard lays down the following rules: (1) general blood-letting is called for when headache continues after
labor is completed, and is attended by flushed face, restlessness
and convulsions of atonic character. Particularly is this procedure demanded when there has not been much loss of blood
at the birth of the child; (2) that the infusion of digitalis is
useful to steady the heart's action, to allay nervous irritation,
and also as a diuretic when aided by the addition of the bitartrate of potassium; (3) that chloroform ought to be used spar-
ingly; (4) that the continuous action of chloral hydrate is
greater than that of chloroform and that it is less likely to

disturb the brain; (5) that in cases in which there has been great loss of blood, or great prostration attended by nervous exhaustion, dependence may be placed upon hypodermic injections of morphia for controlling the convulsions

AMENORRŒA FROM ANÆMIA.

Dr. A. J. C. Skene aims first at restoring the normal condition of the blood, and second, at re-establishing menstruation. Capricious appetite, coated tongue, and constipation he overcomes by giving at the outset a cathartic pill composed of pil. hydrarg. rubbed up with glycerine, syrup. rhei aromat and carbonate of magnesia. If the constipation is obstinate, the following prescription is used:

℞ Quiniæ sulph.,
 Ferri sulph.. āā ℈ij.
 Ex. colocynth. comp........................... gr. x.
 Ex. belladon...................................... gr. ijss.   M.
Sig.—One pill before each meal.

If advisable, aloës is substituted in place of the colocynth.

Among ferruginous tonics Dr. Skene prefers the chlorate of potassium combined with the tincture of the chloride of iron. He also recommends a small quantity of wine or alcohol after meals.

When amenorrhœa persists after the anæmia has been cured, he gives 5-grain doses of the chloride of ammonium every 3 or 4 hours. He believes that this drug favors the rapid exfoliation of the epithelial lining of the endometrium. Dr. Skene is opposed to the employment of emmenogogues in this condition, holding that they almost all act by producing irritation and congestion of the pelvic organs.

RETROFLEXION OF THE UTERUS WITH HYPERÆMIA.

One of Dr. Bozeman's patients had been subject to uterine

hemorrhage more or less severe and frequent, and the fundus had been in such a hyperæmic condition, that the slightest touch of the probe was followed by quite a free flow of blood. This engorgement was treated with applications of a solution of carbolic acid in glycerine, of the strength of half a drachm to the ounce, and also by the frequent use of hot vaginal douches. Under this course of treatment, decided improvement had taken place, notwithstanding the fact that it was as yet impossible to restore the uterus to its normal position. A little later it was proposed to accomplish this restoration by gradual pressure upon the vagina and the fundus uteri by means of the persistent use, for a sufficient length of time, of cotton columns applied in the vagina, in accordance with the method that has proved so successful in Dr. Bozeman's hands.

The plan is original with him, and is somewhat as follows: The patient having been placed in the knee-elbow position, and Bozeman's speculum introduced, a pledget of carbolized cotton is pushed up against the fundus with a pair of dressing-forceps, and held in position there by means of the perineal elevator ordinarily employed in connection with this speculum. A second and third pledget is then applied in the same manner, the perineal elevator being drawn a little further out as each is introduced; and this process is carried on until a firm column of cotton, not stuffing up the whole vagina, but of comparatively narrow diameter, has been formed that reaches obliquely from the fundus of the uterus to the symphysis pubis, which is here the *point d'appui*. Such a column may ordinarily be left in position for about 2 days, but is not allowed to remain for longer than 48 hours. These columns are put in about every three days, the patient being allowed to rest for 24 hours after the removal of each one, and vaginal douches being used in the interval. When by this means the uterus has been restored to its normal position and the vagina to its normal condition, any appropriate support may be worn by the patient as long as is necessary.

TREATMENT OF TYMPANITES.

In some instances, Dr. T. G. Thomas finds systematic kneading of the abdomen to be of service, and that the knee-chest position will often enable the patient to get rid of a considerable quantity of the gas. There is one case which he mentions particularly, in which such marked tympanites came on after ovariotomy that the patient nearly died in consequence of it. When this method was resorted to, there was an escape of an unlimited amount of flatus by the anus, and the patient afterwards made a good recovery from the operation.

CHRONIC OVARITIS.

Dr. T. G. Thomas holds that the patient needs feeding up to the greatest possible extent, in addition to a course of appropriate tonics. One of the best of these, in this condition, is the syrup of the hypophosphites, which is now so frequently employed in the incipient stages of phthisis. The remedies employed are changed from time to time. He has found no other treatment for chronic ovaritis nearly so good as change of air and scene, and the pleasurable excitement of sight-seeing and travel in cheerful company, although he does not pretend to explain exactly in what manner the good result is brought about.

But if the patient is not able to travel in Europe, she is directed to make use of very copious hot water vaginal injections at least twice a day, and three times, if possible. Then the roof of the pelvis is painted once a week with compound tincture of iodine. Iodine is also applied externally, and as often as two or three times a week. At the same time, electricity is faithfully employed, and for this purpose, Dr. Thomas has found the constant current the only one that is of service, the faradic current being rather injurious than beneficial. The best way to apply it, he thinks, is for the patient to lie upon one electrode while the other is carried to the region of the affected ovary, and the application may be made once or twice a week. For at least a week preceding menstruation absolute rest in bed is insisted upon, (rest being as important to an inflamed ovary as to an inflamed eye;) but she is permitted to get up as soon as the flow makes its appearance.

RETROFLEXION WITH FISSURES AND STRICTURES OF THE RECTUM.

The proper treatment in these cases, according to Dr. M. A. Pallen, consists in keeping the bowels well open by enemata. In addition to these, the woman is daily put in the genu-pectoral position and her uterus thrown well forward until it gets into a position in which it can be retained by a pessary. Such patients as these are naturally and habitually constipated, the toleration of the rectum being really wonderful.

Dr. Pallen teaches that another very excellent treatment of these cases is by filling the rectum with very large quantities of hot water, (as the bowel can be educated to be very tolerant,) thereby unfolding the rugæ, or rather expanding them, so that the very bottom of the rugous fissures can be washed out, and any mucosities or pus cleansed therefrom. After the gut is thoroughly washed and the mucous membrane freed from hypersecretion, he considers it proper to touch the eroded or ulcerated spots with nitric acid, nitrate of silver, or even the milder non-caustic astringents. This he easily accomplishes by dilating the rectum with the Sims speculum. But all treatment is useless, if the physician fails to overcome the cause, *i. e.*, to get rid of retroflexion, for the fundus of the uterus jams the anterior upon the posterior wall of the rectum, as the hyperæmia produced by this, as well as the accumulated fæces above, gives rise to rectal catarrh. Dr. Pallen claims that in many cases of this character, an apparent stricture ensues which is in reality no stricture at all, but is a symptom of obstruction, giving rise to the very distressing conditions of the rectal tenesmus and dysenteric discharges.

LEUCORRHŒA.

Due to Fungoid Growths on the Endrometrium.—Dr. T. G. Thomas treats this condition by passing a curette up to the fundus, either after or before dilatation of the cervix, and drawing it gently over both walls of the uterus. After this he keeps the patient quietly in bed for from 2 to 3 days, watching for the occurrence of pain, or increase of temperature. He then supports the womb by the introduction of a pessary, and administers either viscum album in the form of a fluid extract, or 20-drop doses of Squibb's ergot 3 times a day.

Due to Insufficient Diet and Consequent Nervous Depression.—Dr. Thomas insists upon it that such patients eat fresh meat 3 times a day, together with other food, and that they take a tumblerful of fresh milk between meals. At the same time iron, bitter tonics, and beer or ale are ordered.

Due to the Presence of a Cervical Polypus.—The leucorrhœa will disappear as soon as the polypus is snipped off with the scissors. Dr. Thomas lays great stress upon surgery as an element of gynæcology.

Due to Ectropion of the Lining Membrane of the Womb.—Dr. Thomas cures these cases by snipping this ectropion on both sides and turning in the edges of the mucous membrane. He keeps the vaginal walls contracted by the use of astringent vaginal injections.

CHRONIC UTERINE CATARRH.

The first thing done by Dr. Thomas is to put the uterus in its proper position if it be displaced. It is then kept in place by means of an anteversion pessary, and vaginal injections are employed constantly to keep the parts free from irritation. After the next succeeding menstrual period the endometrium is carefully and thoroughly scraped with the curette and all the fungoid growths removed.

CARCINOMA UTERI.

Dr. Thomas, in cases which are destined to be fatal, aims (1) at controlling the hemorrhage, (2) at relieving the pain, and (3) at disinfecting the offensive discharges. The plan which he pursues is as follows: first, he rapidly cleanses the vagina and uterine cavity by means of absorbent cotton, and then applies chemically-pure nitric acid to all the diseased surface. This generally controls the hemorrhage for a longer or shorter period.

For the relief of pain his treatment is all summed up in one word, and that is *opium*. If it does not suit the patient's stomach it is given by the rectum or hypodermically.

He overcomes the disagreeable odor of the discharges by means of very copious vaginal injections given 3 times a day, and consisting of water containing a sufficient amount of thymol or carbolic acid to act as a disinfectant, and some such simple astringent as alum or sulphate of zinc.

As much food is required in such cases as can possibly be digested by the patient. Dr. Thomas considers iron and the hypophosphites as utterly useless. He regards milk as the best article of food, 6 ounces being given at first every 3 hours and afterwards every 2 hours. As soon as practicable a large amount of cream is added, so that the patient may take 2 ounces of cream to every 4 ounces of milk. In this way he has been able to prolong life for a considerable period in several instances.

PUERPERAL PERITONITIS.

The opium treatment is employed by Dr. Clark in cases

where peritonitis is the most prominent element. In Bellevue Hospital five out of six are cured by this treatment. Besides the opium, his patients take a few doses of veratrum viride to diminish the frequency of pulse. He gives Norwood's mixture of veratrum in doses of gtt. v, when the opium has reduced respiration but not the pulse. It sometimes produces great nausea, attended by prostration and a tendency to syncope. Alcoholics are used when such effects are produced. He considers it a very good treatment to give opium and veratrum viride in alternate doses, and regards this as all that is necessary. In *metro-peritonitis* opium does not serve any important purpose, and he thinks it useless to give it, except to *soothe* the patient. Leeches to the vulva or perinæum and bleeding are very necessary. He very often employs injections of warm and tepid water into vagina and uterus. During the period of purulent infection, he prescribes quinia sulph., (gr. xv per day,) combined with morph. sulph., to reduce irritability. If there is a tendency to the formation of abscesses, food and stimulants are, of course, necessary.

VESICO-UTERINE FISTULA.

Dr. Bozeman's operation for the cure of this condition, consists in the complete excision of the anterior lip of the cervix uteri, together with a part of the vesico-vaginal septum, thus converting the original lesion into a vesico-utero vaginal fistula. This being done, the posterior lip of the opening, which was the stump of the cervix uteri, is next pared off, as is generally done in fistulæ of the latter class. Four silver-wire sutures are needed, which are introduced by a straight needle set in a curved needle-holder. This being done, the sutures are next adjusted in the usual manner, and a button or plate of lead of suitable form and size, is slid down upon them, and the whole then secured in place by the compression upon each wire of a perforated shot. The great utility claimed for this form of suture in the operation centres in the leaden plate, which stands across the cervical canal and prevents its recontraction and the consequent puckering of the line of coaptated edges, until union takes place. After washing out the bladder, the patient is placed in bed and quiniæ sulph. gr. x, and liq. opii comp. f ʒ j., administered per rectum, and followed by gr. j of opium by the mouth, every six hours. Dr. Bozeman claims great advantage in the operation from the use of the self-sustaining and dilating speculum. By means of it the vagina is expanded to the fullest extent, and the greatest facility afforded to the movements of instruments. The advantages of the knee-chest position are also fully illustrated. The patient, resting upon a supporting apparatus, takes the anæsthetic with the greatest comfort.

NERVOUS DISEASES.

SPERMATORRHŒA.

Constipation.—An enema of cold water is ordered every morning. This not only produces a normal evacuation, but also stimulates the blood-vessels and the surrounding parts to a more vigorous contraction and accelerates their return to a normal condition.

Derangement of Digestion.—A diet is recommended by Dr. Joseph W. Howe calculated to increase the patient's vitality, and consisting of oysters, eggs, milk, beef, mutton, etc.

Stimulants.—Some mild wine, such as claret, is prescribed, which will promote good digestion without exciting inordinate desires.

Bathing.—The patient is directed to take a cold sponge-bath every morning. Cold water is, at the same time, ordered to be thrown into the rectum.

Exercise of all kinds is advised, with the exception of horseback riding.

Local Treatment.—Dr. Howe does not recommend caustics or the passage of the sound, but thinks very highly of electricity. One electrode is insulated to nearly its entire extent, except that part which rests against the prostatic urethra. The other electrode is applied over the fourth lumbar vertebra. Only a very feeble current is allowed to pass through at first, and the first sitting only lasts five minutes. The second day, the wire-brush is used, passing it over the inside of the thighs, about the perinæum. On the third day, the urethral electrode is again employed. The strength of the current is gradually increased with each sitting.

Medical Treatment.—One of the best tonics, in Dr. Howe's opinion, is this:

℞ Strych. sulph.. gr. j.
    Quin. sulph... ʒ ss.
    Tc. ferri mur... f ʒ ss.
    Glycerinæ... f ʒ iv.   M.

SIG.—One-half teaspoonful in a wineglass of water, four times a day half an hour before meals, and at bedtime.

This is another favorite prescription:

℞ Ferri arsen.,
    Ex. nuc. vom...................................... āā gr. v
    Ergot.,
    Quin. sulph.. āā ʒ ss.   M.
        Et in pil. No. xxx div.
SIG.—One pill four times a day.

Where constipation is a prominent symptom, gr. x of aloës are substituted for ergot, in the above.

Impotence:

℞ Ex nuc. vom.. gr. ¼.
    Phosphor.. gr. 1/100.   M.
SIG.—To be taken after meals.

In some cases, f ʒ ss doses of the fluid extract of damiana are substituted for the phosphorus, with advantage, or, another excellent remedy, f ʒ ss–j of the tincture of water pepper. Here are three prescriptions which he frequently uses:

1. ℞ Tc. canthar.,
        Tc. nuc. vom.,
        Tr. ergot .... āā f ʒ j.   M.
    SIG.—Ten to twenty drops, four times daily.

2. ℞ Tr. sanguinariæ................................... f ʒ ss.
        Fl. ex. stillingiæ................................ f ʒ ij.   M.
    SIG.—Twenty to thirty drops, four times a day.

3. ℞ Capsici.. gr. x.
        Quin. sulph... gr. v.
        Vini xerici.. f ʒ jss.   M.
    SIG.—To be taken at bedtime.

Over-Excitement of the Genital Organs.—This is controlled by gr. xx of bromide of potassium at night, and four times a week. During the second week, the dose is increased to gr. xxx. The use of the bromide is always preceded by a brisk cathartic. According to Dr. Howe, the radical cure of this class of cases consists in the patient's getting married.

TUBERCULAR MENINGITIS IN CHILDREN.

Dr. Delafield does not care to disturb the child by applying blisters to the nape of the neck, etc. He quiets the vomiting and increases the amount of urine secreted by the kidneys by administering the bicarbonate of potassium and lemon juice. In order to still further increase the secretion of urine, the child is placed in a tub of warm water for about five minutes, then removed, and, without drying the skin, wrapped in warm blankets and allowed to sweat for two or three hours. When something is required to keep the patient quiet, a mixture of chloral hydrate and bromide of potassium is given.

MIGRAINE.

Dr. Eugene Depuy has obtained good results in some cases by introducing brandy or strong snuff into the nostril upon the side corresponding to that upon which the pain is felt. Another excellent method of treatment is by throwing carbonic acid gas against the nasal mucous membrane in a sufficiently strong jet to produce a marked impression. In still other cases, Dr. Depuy has administered the so-called "potion of Riverius," which consists of citric acid and simple syrup, and bicarbonate of potassium and water, adding to it bromide of potassium. This is an effervescing mixture, and is administered in alternate tablespoonfuls, allowing effervescence to take place in the stomach. He recommends also occasional mustard plasters to the nape of the neck, and the use of mild aloëtic aperients.

Dr. E. C. Seguin recommends that when a patient wakes in the morning with a feeling as if a headache were imminent, a drachm of paullinia powder, or the same quantity of its equivalent, the elixir or fluid extract of guarana, be taken, the dose to be repeated in the course of an hour, unless relieved. When the pain is severe, he does not hesitate to use a hypodermic of $1/100$ of a grain of atropia, combined with from 5 to 15 minims of Magendie's solution.

In the intervals between the attacks he administers ½ a grain of the solid extract of cannabis indica, daily—this dose being kept up for a long time continuously.

Dr. G. M. Beard has had very good results from the administration of caffeine, just before the attack, and of from 15 to 20 grains of the muriate of ammonia.

TRIGEMINAL NEURALGIAS.

Dr. E. C. Seguin has obtained invariably good results from the use of Duquesnel's aconitia. His average dose is $1/100$ of a grain every four hours. He has found that the susceptibility of individuals to this preparation varies exceedingly. Some are over-affected by the $1/200$ of a grain, while others take the $1/84$ with impunity. He concludes that Duquesnel's aconitia is the most powerful and the best remedy for the relief and cure of trigeminal neuralgia.

CEREBRO-SPINAL MENINGITIS.

Dr. Thomson administers teaspoonful doses of the fluid extract of ergot and gr. v of quinia every three hours. One-twentieth of a grain of calomel is given every half-hour. Leeches are applied to the spine, and ice to the head and back. If the pulse is feeble, f℥ iv of whiskey are given every half hour.

The following is sometimes useful:

℞ Potasii iod... gr. xl.
   Ex. conii fl... gtt. xl.
   Aquæ.. q. s. ad f℥ ij.   M.
   Sig.—Two fluid drachms thrice daily.

HEMIPLEGIA.

If of syphilitic origin Dr. Thomson gives the iodide of potassium in doses of a drachm and upwards. He also recommends:

℞ Acid. phos. dil... f℥ vj.
   Syrup. hypophos.............................. q. s. ad f℥ iv.   M.
   Sig.—Two teaspoonfuls in water thrice daily.

The patient is placed as much as possible upon a vegetable and fruit diet. Meat is forbidden, and milk, eggs and fish allowed only in small quantities. Tea and coffee are taken but once a day; all alcoholics, especially malt liquors, are forbidden. Moderate exercise in the fresh air is advised, but fatigue is always to be avoided. Iron is given when muscular degeneration is feared, and corrosive sublimate prescribed in doses of $1/30$ grain, thrice daily for a long period of time.

CEREBRAL HEMORRHAGE.

Dr. Hamilton meets increase in the frequency of the pulse and elevation in temperature, by local derivatives and cardiac sedatives. He thinks it unwise to use electricity in any form, if degeneration has begun. If there is pain, the actual cautery is applied over the nerve trunks. He has derived the greatest benefit from wrapping the limbs carefully with cotton batting, and covering it with oil silk. Tremor is controlled by conium and the avoidance of general excitement. Hot baths and soaking the limbs for 10 or 15 minutes, daily, in water as warm as the patient can bear it, are useful.

PROGRESSIVE MUSCULAR ATROPHY.

Dr. Hamilton employs electricity, and, if the extensors are atrophied, gives support to the hand by means of the rubber muscle.

POLIO-MYELITIS

Is treated by electricity, with a minimum galvanic current, and gradually increasing its strength, allowing the muscles to rest for a day or so between each séance. In addition to electricity, Dr. Hamilton has employed cod-liver oil, the syrup of the iodide of iron, and strychnia, with advantage.

SCLEROSIS OF THE SPINAL CORD.

In the early stages, Dr. A. McLane Hamilton uses ergot with good results. Later, he relies mainly upon phosphorus and cod-liver oil, joined with galvanization and cauterization of the cord. For the relief of the pains, he finds that hypodermic injections of atrophia, morphia, or muscarine, act most favorably. For the same purpose he employs the galvanic current, placing the positive pole over the painful point in the back. Warm sulphur baths, made by simply dissolving an ounce or so of the sulphuret of potassium in water, of a temperature not exceeding 90° F., are useful adjuvants.

EPILEPSY.

Dr. A. McL. Hamilton insists upon a careful observance of hygienic rules. As regards medicine, he is in the habit of combining the bromide of sodium with equal parts of the bromide of ammonium, and of administering ℥j of the combined salts, daily, together with gr. xxx of the hydrate of chloral. The doses are divided so that the largest is given just before the fit is expected. In other cases, Brown-

Séquard's mixture of the bromides with bicarbonate of potassium and a bitter tonic, acts admirably.

Dr. Hamilton regards it as of the utmost importance to combine cod-liver oil, cream, extract of malt, or linseed oil, with the bromides in the treatment.

Where the disease has no specific cause, he resorts to the use of the actual cautery, or applies repeated blisters to the back of the neck. He believes curare to be indicated in obstinate cases, and injects a standard solution of this drug, acidulated with diluted hydrochloric acid, hypodermically, every fifth day, in doses of ⅓ of a grain, until five or six doses are given. In the lighter forms of the disease, he employs f℥j doses of the fluid extract of ergot, thrice daily, alternated with gtt. v doses of tincture of belladonna, the quantity being gradually increased.

When the case gives a specific history, he combines the iodide of potassium, or better still, the bichloride of mercury, with the bromides, pushing the administration of the former drugs as far as he can with safety.

Dr. E. C. Seguin treats idiopathic epilepsy with the following formulæ:

℞ Potassii brom... ℥j.
   Ammon. brom... ℥ss.
   Aquæ font... f℥vij.   M.
Sig.—To be given by the teaspoonful.

℞ Sodii brom... ℥j.
   Ammon. brom... ℥ss.
   Aquæ font... f℥vij.
Sig.—To be given by the teaspoonful.

The quantity administered is so divided as to give the largest dose in the evening. The dose is gradually increased to the production of bromism. The dose is administered in a tumblerful of water. The bromides are continued for at least three years after the last attack. The acne consequent upon the long-continued use of the bromides, is combated with arsenic, sulphur ointments, mercurial plaster, and alkaline lotions. To correct the debility and paresis, strychnia, nux vomica, the oxide of zinc, and quinia, are given. Nitrate of amyl and stimulants relieve the dizziness. The patient's diet is regulated. Cream, cod-liver oil, iron, quinia, phosphorus, strychnia with nitro-muriatic acid, wine, beer, and whiskey, are taken steadily, as tonics and nutrients.

INSANITY.

The bromides are employed by Dr. Seguin to meet such indications as epileptiform attacks, or abnormal sexual excitement, or great nervousness not caused by delusions.

FACIAL NEURALGIA

Is treated at the Presbyterian Hospital by croton chloral given in solution with elixir calisaya; 5 grains 3 times a day for a week.

INSOMNIA.

Dr. E. C. Seguin prescribes chloral in some cases, and in others some stimulant, such as beer. He does not place much confidence upon the use of the bromides here.

HAY-ASTHMA.

Dr. Seguin uses this gargle:

R Ammon. brom., (ℨj to ℨij-fℨj)..................... fℨvj. M.
Sig.—To be used as a gargle.

The nasal passages are washed out several times a week with a weak solution of the same salt, (gr. x–xxx to fℨj.)

CHOREA.

If the child is run down, Dr. Hamilton administers iron and cod-liver oil. Strychnia is given up to the point of producing stiffness of the sural muscles. He thinks that the application of cold to the spine cannot be over-estimated as a plan of treatment. He either employs the ether spray, or applies ice-bags, allowing them to stay on about ten minutes.

The spray is directed to the upper part of the cord, over the upper cervical vertebræ. Eserine he regards as a dangerous remedy and one likely to produce severe gastric symptoms. Where nothing else does good, Dr. Hamilton is accustomed to put his patients in a dark room and keep them quiet. The diet is carefully regulated. Among useful hygienic measures are the salt bath and the energetic use of the rough towel.

SPINAL ANÆMIA.

Dr. Hammond prescribes gr. $1/10$ of the phosphide of zinc with gr. $\frac{1}{2}$ of the extract of nux vomica, in pill form, to be taken three times a day. Lately he has pursued the practice of giving strychnia in gradually increasing doses, until there is some evidence of the production of its characteristic physical effects. He dissolves gr. ij of the sulphate of strychnia in f℥j of water, and gives ℳx (containing gr. $1/24$ of strychnia) three times during the day. On the next day, ℳxj are given at each dose, and on the third day ℳxij, and so on until the paralysis yields, or the muscles of the legs become stiff. In this latter case, the use of the drug is stopped for a day, and on the next day he begins again with the original dose.

ACUTE CEREBRAL MENINGITIS.

If the cephalalgia is intense, Dr. Hammond takes as much as 12 or 16 ounces of blood from the arm. Leeches are applied behind the ears. The hair is cut off short, and ice kept constantly applied to the scalp. As a purgative, gr. x of calomel, and gr. ij of podophyllin, are administered. He has derived the greatest benefit from the bromide of potassium, in 30-gr. doses three or four times a day. The head is kept well elevated and care is taken to keep the room in which the patient lies, cool and well ventilated, and to exclude the light as much as possible. As food, the chief reliance is placed upon strong beef tea. When the strength flags in the later stages of the disease, alcohol is administered in appropriate doses.

HYOSCYAMIA AS AN HYPNOTIC AND ANTISPASMODIC.

Dr. E. C. Seguin reaches the following *provisional conclusions* with regard to the modes of action of this drug:

1. It acts upon the pupil as a mydriatic.
2. It reduces the pulse gradually and increases arterial tension.
3. It checks bodily heat.
4. It produces hallucinations and delirium.
5. Its use is occasionally attended by a rash.
6. In large doses, it produces sleep, and something like paralysis or paresis, and may induce retention and dysuria.
7. Theoretically, it is indicated in mania attended by restlessness, delusions and suspicions, and in insomnia and convulsive affections.
8. It has been of especial service in acute or subacute mania, insomnia, and those cases characterized by mischievous delirium.
9. It induces sleep more certainly than chloral, and without being followed by bad effects.
10. In paralysis agitans it can do what no other remedy can do.
11. It is a diuretic of no mean power.
12. Its curative power does not seem to be great.
13. In acute chorea its use may play an important part.

Mode of Administration and Proper Dose.—It can be given

in small doses, hypodermically, with ease. The doses are from $1/20$ to 1 grain of the amorphous, and from $1/100$ to $1/25$ of a grain for hypodermic use. Distinct effects may be obtained from the $1/100$ of a grain.

The following formula is given for hypodermic use:

℞ Hyoscyamiæ (Merck's cryst.).................... gr. j.
Glycerinæ,
Aquæ.. āā ℳ 100.
Acid. carb. puræ.................................. gtt. j.   M.

Each minim contains $1/200$ of a grain.

Tablets containing $1/50$ of a grain are convenient for use by the mouth.

CEREBRAL ANÆMIA.

Dr. Hammond regards the alcoholics as the indications *par excellence*. The quantity employed is small at first and is administered frequently and in a highly diluted form. When alcohol cannot be employed, the carbonate or aromatic spirits of ammonia are given.

As a tonic he regards this mixture as of great value:

℞ Strychniæ sulph................................ gr. j.
Ferri pyrophos.,
Quiniæ sulph.................................... āā ʒj.
Acid. phos. dil.,
Syr. zingiber................................... āā f℥ ij.   M.
Ft. mist.

SIG.—A teaspoonful three times a day in a little water.

The diet used must be of good quality, consisting chiefly of milk, eggs, and meat of various kinds.

The patient is encouraged to pass a good portion of each day in a recumbent position. Emotional disturbance and anything above moderate mental exercise is to be sedulously avoided.

Bromide of potassium is strongly counter-indicated.

SPINAL CONGESTION.

In cases which come on suddenly, Dr. Hammond draws blood from the spine by cups or leeches. He thinks the verge of the anus the best place for the application of the latter. As a purgative he administers ℨj doses of the sulphate of magnesia 2 or 3 times a day. He regards the ergot of rye as of great value in this disorder, and gives it in fℨj doses of the fluid extract 3 times a day. When there is paralysis of the sphincter of the bladder, or when the pain in the back is severe, gtt. xv doses of the tincture of belladonna 3 times a day, and a belladonna plaster is applied to the painful region of the spine. He has found the hot douche to be an excellent means of drawing the blood from the deep to the superficial vessels of the spine. The water is allowed to fall upon the naked back over the diseased part of the cord every day for about 5 minutes. Dry cups have also been found to be valuable adjuncts. The constant current is applied to the spine over the affected part of the cord, the positive pole being held at the upper limit of the lesion and the negative pole rubbed up and down, over all the parts below. The *séances* should not exceed 10 minutes in length. As an application to the paralyzed muscles the induced current is employed, being used every day for half an hour or so at a time.

Dr. Hammond considers that phosphorus and strychnia should never be given in cases of spinal congestion.

INDEX.

A.

Abortion..................................81–82
Abscess................................... 64
Acne Rosacea.......................21–22
Alimentation, Rectal................ 18
Amenorrhœa from Anæmia....... 86
Ammonia, Intravenous Injection of.. 17
Anæmia, Amenorrhœa from...... 86
" , Cerebral.................... 104
" , Spinal 102
Anchylosis, Fibrous................59–60
Angina Pectoris....................... 51
Ano, Fistula in......................... 62
Anthrax.................................... 64
Aortic Lesions......................... 51
Asthma..................................... 6
" , Hay 101

B.

Boracic Acid as an Anodyne..... 63
Bright's Disease, Chronic 31–32 and 47
Burns....................................... 64

C.

Cancer of Tongue, Ligation of Sublingual Artery for Removal of..................................58–59
Catarrh, Chronic Uterine.........90–91
" , Naso-Pharyngeal......29–31
Cervix Uteri, Laceration of...... 79
Chancre 68–70
Cholera Infantum..................39–40
Chorea 102
Colles' Fracture....................66–67

Cough Mixture, A Disinfectant and Antiseptic...................... 53
Congestion, Spinal................... 105
Constipation associated with Chlorosis 49–50
Constipation dependent upon deficient Secretion.................... 48
Constipation due to Chronic Inflammation of the Rectum..... 49
Constipation due to want of Peristaltic Action..................... 49
Constipation following Febrile Diseases............................. 49
Contusions 51
Convulsions, Puerperal..........85–86
Cystitis following Gonorrhœa...73–74
Cystorrhagia.........................74–75

D.

Diabetes Insipidus in Children... 53
Diarrhœa, Beef Tea in............. 32
" , Chronic, in Adults...15–17
" due to Errors in Diet, 36–37
" due to Inflammatory Disorders............................. 35
Diarrhœa due to Preternatural Acidity................................. 37
Diarrhœa, Dysenteric.............33–34
" , Flatulent 34
" , General Hints in....38–39
" , Koumyss in...........34–35
" , The, of Dentition...37–38
" , The Reduction of Temperature in...................32–33
Diphtheria............................. 5–6
Dysentery, Acute....................14–15

INDEX.

Dysentery, Chronic.................. 15
" , Epidemic............... 15

E.

Ear, Accumulations in the........ 19
" , Catarrhal Inflammation of Middle............................46-47
Ear, Chronic Suppurative Inflammation of Middle............ 7-9
Eczema Rubra....................... 66
Emesis................................. 19
Emphysema 52-53
Empyema...............3-4 and 62-63
Enteritis, Subacute.................. 15
Epididymitis......................67-68
Epilepsy.........................99-101
Epistaxis.............................. 53

F.

Favus................................... 50
Femur, Fractures of................ 67
Fever, Scarlet.....................51-52
" , Typhoid...................22-23
" , " in Children....40-42
Fibula, Fracture of.................. 67
Fistula, Perineal..................... 68
" , Vesico-Uterine...........92-93

G.

Gonorrhœa, Cystitis following..73-74

H.

Hæmoptysis 14
Hæmorrhoids, Internal...........64-66
" , Treatment of....71-72
Headache dependent on Gout... 43
" , Dyspeptic.............44-45
" of Acute Alcoholism, 44
" " " Cerebral Congestion........................45-46
Headache of Passive Cerebral Congestion........................ 46
Headache of Cerebral Anæmia.. 46

Headache of Cerebral Effusion... 44
" " " Softening, 46
" " " Tumors... 46
" " Rheumatism......... 43
" " Syphilis.............. 43
" , Malarial................. 43
" , Nervous................. 42
" , Sick....................42-43
" , Uræmic.................43-44
Heart, Disease of the............50-51
Hemiplegia............................ 98
Hemorrhage, Cerebral............. 98
" , Post-Partum....... 78
Hernia, Stangulated and Incarcerated............................57-58
Hip-joint, Luxation at the......54-55
Hyoscyamia as an Hypnotic and Antispasmodic...............103-104

I.

Insanity............................... 101
Insomnia............................. 101

J.

Jaundice 11

L.

Laryngitis, Catarrhal............... 48
Leprosy, The Larynx of.......... 31
Leucorrhœa due to Ectropion of the Lining Membrane of the Womb.............................. 90
Leucorrhœa due to Fungoid Growths on the Endometrium 90
Leucorrhœa due to Insufficient Diet and Consequent Nervous Depression 90
Leucorrhœa due to the Pressure of a Cervical Polypus........... 90
Lingual Artery, Ligation of, in Removal of Cancer of the Tongue............................58-59
Lip, Cancer of the Lower.......60-61

Liver, Cirrhosis of the............ 20
Locomotor Ataxia, Treatment
  of the Pains of..................12-13
Lungs, Gangrene of................. 13
  ", Multiple Abscess of...... 13

M.

Mamma, Scirrhus of the.......... 59
Meningitis, Acute Cerebral...102-103
  ", Cerebro-Spinal......97-98
  ", Tubercular, in Children................................. 96
Menorrhagia......................77-78
Migraine 96-97
Morphia Vomiting.................... 19
Muscular Atrophy, Progressive, 99

N.

Nævus................................ 55
Nephritis, Chronic.................. 20
Neuralgia, Facial................... 101
  ", Trigeminal............. 97
Nipples, Sore........................ 83

O.

Opium Poisoning..................... 19
Ophthalmia Neonatorum......... 17
Otorrhœa, Acute Primary.......17-18
Ovariotomy...........................83-85
Ovaritis, Chronic...................88-89

P.

Patella, Fractured................... 67
Pelves, Labor in Kyphotic....... 77
Perineum, Laceration of the..... 77
Peritonitis, Puerperal............91-92
  ", Sporadic..............11-12
Pleurisy, Acute....................... 9
  ", Subacute.................. 6-7
Pharyngitis, Croupous............. 22
Phthisis, Aphthæ of................ 10
  ", Diarrhœa of............... 10
  ", Laryngeal................28-29
  ", Sore Throat of.........10-11

Phthisis, The Rheumatism of.... 14
  ", Vomiting in.............. 10
Pneumonia, Acute................... 1-3
  ", Broncho..............20-21
  ", Chronic.............13-14
  ", Lobular................ 13
Polio-Myelitis........................ 99
Pregnancy, Abdominal............ 79
Pruritus Vulvæ due to Vaginal
  Leucorrhœa.....................78-79

R.

Rectum, Extirpation of for Cancer.................................... 59
Rectum, Fissure of the........... 54
  ", Prolapse of the.......... 58
  ", Retroflexion with Fissures and Stricture of the.....89-90
Regurgitation, Mitral and Aortic...................................18-19
Retroflexion with Fissures and
  Stricture of Rectum...........89-90
Rheumatism, Acute................. 19

S.

Skin, Transplantation of the..... 68
Silicate of Lime Splint............ 63
Spermatorrhœa 94-96
Spinal Cord, Sclerosis of the..... 99
Spine, Potts' Disease of the.....55-56
Stomach, Rum........................ 52
Strapping, Wet....................... 66
Stricture 56-57
Surgical Dressing.................... 63
Synovitis, Chronic.................. 60

T.

Tibia, Fracture of the............. 67
Toe-Nails, Ingrowing.............. 61
Tracoma, Acute...................... 28
Tympanites, Treatment of........ 88

U.

Ulcer, The Chancroidal.........70-71

Uræmia.................................. 20
Urethritis, Acute, Non-Specific, 68
Urine, Acute Retention of the, 72–73
Uteri, Carcinoma..................... 91
Uterus, Retroflexion of, with Hyperæmia.......................86–87

V.

Vaginal Leucorrhœa, Pruritus
Vulvæ due to.....................78–79
Valvular Lesions..................50–51
Vaginismus........................... 76

W.

Womb, Submucous and Interstitial Fibroids of...............79–81
Wrist-joint, Inflammation of the, 62

NOTES

OF

HOSPITAL PRACTICE.

PART III.

NEW YORK AND PHILADELPHIA HOSPITALS.

EDITED BY

SAMUEL M. MILLER, M.D.

PHILADELPHIA, PA.:
SAMUEL M. MILLER, Publisher.
1881.

GENERAL DISEASES.

ACUTE RHEUMATISM.

Dr. Austin Flint says :—There is an important point in practical medicine to which I wish to direct attention, and it consists in the use of salicylic acid as an anti-rheumatic remedy. It seems to me that it should not supersede the alkaline treatment which has been employed to diminish the liability to cardiac complication. It has not as yet been proved that salicylic acid has any effect in the way of preventing cardiac complications except by way of shortening the duration of the rheumatic fever. I have had occasion to observe several cases of pericarditis occurring in the course of cases of articular rheumatism under treatment by the use of salicylic acid exclusively. Because a remedy has been found that apparently causes the disease to abort occasionally, or, if not that, shortens its duration, we are not to relinquish the accepted alkaline treatment, but should carry it to its full extent as we have been accustomed to do heretofore. The alkaline treatment does not exert a marked effect upon the duration of the disease ; but the weight of evidence showing that it diminishes the liability to pericarditis and endocarditis is overwhelming. Fortunately, the two plans of treatment do not conflict with each other.

PORTAL THROMBOSIS.

The only remedy which offers any prospect of relief is ammonia, which has the power to dissolve coagula. Unfortunately, the stasis in the portal system so hinders absorption that remedies do not readily enter the blood. As Halfourd, of Australia, has demonstrated the innocuousness of the intravenous injection of ammonia, this expedient according to Dr. Barthalow should be practiced in such cases. It consists in the injection of one part of aqua ammoniæ to two parts of water into any convenient vein. If, however, there be any movement of blood in the portal, the ammonia should be administered in the form of the carbonate—five grains every three hours. The usual remedies for ascites will be necessary.

PULMONARY HÆMORRAGE.

The very best remedy, in Dr. Barthalow's opinion, is Squibb's extract of ergot, or ergotine, as it is termed. Of this, as much as twenty or thirty grains may be given hypodermically, although so large a quantity is seldom necessary, and it will generally suffice to introduce five or six grains. It is exceedingly difficult to decide whether any remedy is efficient in this condition, since the hæmorrhage constantly subsides spontaneously, and any drug that happens to be given at the time, of course, gets the credit; but the stopping of the spitting of blood so often follows the injection of ergot that he has no doubt that these cases are benefited by its administration. Another remedy is ipecac. It seem strange to use it in pulmonary hæmorrhage, but he thinks it is one of the best means that we have. In causing nausea and vomiting it affects directly the pulmonary circulation. You should give enough ipecac to cause nausea, and be indifferent whether it causes vomiting or not. One of the dangers of the condition is that the blood will remain in the air cells and smaller tubes and close them, and thus set up irritation and further mischief. The administration of ipecac has the advantage of clearing the lobules and at the same time it has an influence upon the circulation, which makes the vomiting entirely safe. He uses this formula.

℞ Extract. ergot. fluid., ... f℥ss.—i.
   Extract. ipecac. fluid., ... ℳv. M.
Sig.—Every three or four hours.

Ice should be applied to the chest, and pieces of ice allowed to melt in the mouth. The patient is to be kept as quiet as possible, in a semi-recumbent posture. A very common household remedy is table salt, and it is not without effect, but ice is more valuable. A large piece of ice placed at the nape of the neck will sometimes succeed, especially if followed by hot water. The quick alternation of heat and cold produces a most decided contraction of the arterioles, and is better than cold alone.

If the hæmorrhage prove persistent, he employs bloodletting, in order to quickly reduce the blood-pressure.

DIABETES MELLITUS.

The dietetic treatment holds the first place in Dr. Flint's opinion. This treatment consists in withholding sugar in any form almost entirely from the food. Not only is this necessary in the case of sugar, but also all the starchy constituents of diet capable of being transformed into sugar. The place of wheaten bread, he thinks, is best supplied by the gluten bread prepared by the Health Food Company. Although Dr. Flint believes that *we have no other resources if the dietetic treatment does not succeed*, yet there is one remedy which he occasionally employs. This is the sulphide of calcium, a fifth of a grain three times a day.

NASAL CATARRH.

The best treatment for nasal catarrh, in Dr. Jacobi's opinion, is summed up in two words—absolute cleanliness. The nose must be regularly washed out with tepid water in which a little salt has been dissolved. The solution is to be a weak one, not stronger than from a half of one per cent. to one per cent. of salt. This solution is to be snuffed up into the nostrils until it can be spit out through the mouth. In this way a tumblerful of the solution is to be used three or four times a day.

This constant washing out prevents the mucous from collecting in the nasal passages and getting rancid. As a rule, cases get well in two or three months. If you wish to make use of some medicinal application, nitrate of silver is the best. Dr. Jacobi uses it almost exclusively. He thinks there is no more dangerous practice than the use of nitrate of silver in stick or in concentrated solution. There are many chronic cases which he has seen in which the mucous membrane presents a peculiar appearance. It is shiny, hard, not moist, and thin. This condition is incurable, and is the frequent result of strong solutions of nitrate of silver. In other cases the effect produced has been caustive, not alterative, and has resulted in a cicatrical condition of the surface to which it has been applied. Where a mild solution is used the effect is altogether different. Under the microscope the fluid may be seen to find its way into the interior between the epithelial cells. It really changes the morbid circulation into a healthy one. The strength of a solution to produce such an effect ought not to be more than from a quarter to two grains of the nitrate to an ounce. The same thing is true of solutions which are meant for the bladder. A great deal of harm can be done by the use of concentrated solutions, and never more good than by mild ones. As a rule Dr. Jacobi uses a solution of one-fifth to one-tenth per cent., that is, one-half to one grain. This he injects twice a week, seeing that it enters the pharynx properly. Then, during the intervals, he lets the patient attend to the salt and water washing himself. Stronger solutions than these give pain. These do not, only creating a slight uneasiness. Such a point as this seems a trifle, but it is a trifle worth remembering. Recollect that it is by attention to trifles you will be able to cure disease. It is not the extraordinary and brilliant operations, the feats of medicine, which insure success in the treatment of disease so much as this strict attention to trifles.

For the pharyngeal catarrh he uses the spray. The tube should be introduced through the nose. When you do that, remember that the floor of the nasal cavity is parallel with the surface of the earth. Then, when the spray has been introduced, let the patient inhale once or twice. When they feel the spray in the mouth you must cease. Then if you look into the throat you will see on the pharyngeal wall a slight, whitish discoloration, nothing more.

INDIGESTION.

In the treatment Dr. Barthalow first removes from the diet the articles to which this acid is due, that is, he takes care to remove the farinaceous articles of food. He also gives ten drops of diluted muriatic acid before meals, not after meals, for then it would only add fuel to the flame. Acid is given before meals to reduce the acidity of the gastric juice; it checks the formation of acid, in accordance with the well-known physical law, if you put an acid upon one side of an animal membrane, and an alkali upon the opposite side, there is rapid diffusion, and thus, availing ourselves of this fact, by giving acid before meals we may reduce the acidity of the gastric juice. An acid applied to the mouth of a follicle, which normally excretes an acid fluid, will reduce the acidity of the glandular secretion; this is a physiological law, as well as a clinical fact. Upon this principle he conducts the management of the case, and with due attention to diet, and to the state of the bowels, expects in a short time to effect a cure.

ACUTE TONSILLITIS.

The first thing Dr. Van Valzah does is to freely puncture and scarify the tonsils and soft palate, thereby favoring free depletion. This gives immediate and marked relief, as indicated by the improvement in the patient's talking, breathing, and swallowing. He cannot too earnestly emphasize the value of free scarification in cases of this kind. It is a simple procedure which at once relieves the dangerous congestion and inflammation, and affords the patient decided comfort. All that is necessary to be done is to take a sharp-pointed instrument, like a tenotome or bistoury, and puncture the parts or make a few superficial incisions. Having depleted the parts thoroughly, he then makes free use of ice, both internally and externally.

It is always proper in this class of cases to insure cathartic action of the bowels, and for that purpose he gives the patient three grains of pil. hydrargyri and one-half grain of resin of podophyllin. After waiting the requisite time and no movement resulting, he gives a wineglass full of citrate of magnesia every hour, until a pint of the solution has been taken. Both of these remedies failing to give the decided action which is necessary, rectal injections of turpentine, soap and water, are employed, after which purgation is established A gargle is then ordered for the throat, composed of a saturated solution of chlorate of potash, and two drachms of tincture of myrrh, in four ounces of infusion of cinchona. With this he is to gargle at first every hour, but as the symptoms subside, every second or third hour. After the acute symptoms have largely disappeared, the patient is placed on a tonic, consisting of twelve drops of the tincture of iron every four hours.

INTERMITTENT FEVER.

Quinine is, of course, relied upon chiefly; thirty grains a day, in three doses, usually proving sufficient to arrest the chills. It is often found that, where the case is one of long standing, good effects are secured by combining capsicum and opium with the quinine. A powder is given containing quiniæ, gr. x.: capsici, gr. vj.; opii, gr. j., about three hours before the chill is due. This powder may be divided into capsules, or may be given in coffee, which disguises the taste as well as anything. If the intermittent fever is of the tertian variety, quinine is given on the odd day, and the powder on the day of the chill. If it is quatidian, quinine is given at night, the powder in the morning. The use of hypodermic injections of pilocarpin during the chill has been employed in many cases. A new remedy has been tried, successfully in some cases, to arrest the paroxysm or to shorten its duration, viz., thirty drops of the spirits of chloroform in a teaspoonful of glycerine. This may be given safely in cases which have no cardiac symptoms, and it has stopped the chill in some instances promptly. In the treatment of intermittent fevers in infants it is often impossible to use sulphate of quinine, as it produces great disorder of the stomach, and the child may refuse to take its nourishment as well as to take the second dose. Some mothers, under these circumstances, will object to forcing the remedy down. In these cases it is recommended that the tannate of quinine be employed in doses double those of the sulphate. It rarely produces any nausea or vomiting, and the taste can be readily concealed by mixing the drug with chocolate. If to half a teaspoonful of chocolate the requisite amount of the tannate be added, and the whole stirred into a little water or milk, the child will not object to the dose. The anæmia of this class of patients is treated with the tincture ferri chloridi, and in children's cases it is found best to add to this Fowler's solution in drop doses. Cod-liver in emulsion is also largely used, and the emulsion which is now employed is made by mixing equal parts of cod-liver oil and lime-water, and adding a small amount of oil of wintergreen to flavor.— *Bellevue Hospital.*

THE COUGH OF PHTHISIS.

The use of oxalate of cerium in the chronic cough of phthisis and chronic bronchitis is meeting with some favor. Since its introduction into Bellevue Hospital, in May last, a large number of cases have been treated, but as yet no positive indications can be given for its use. It is given in doses of 5 to 10 grains, the usual course being to begin with 5 grains and gradually increase until the patient is taking 10 grains three times daily. It is given in the form of a powder. It seldom disagrees with the stomach, and in a fair proportion of cases is of decided benefit, quieting the distressing cough, and enabling the patient to obtain the needed sleep. It is suggested that it acts as a sedative to the hyperæsthetic nerves of the bronchi, just as it is supposed to arrest the nervous vomiting of pregnancy, but observations are needed to explain fully its method of action. As a matter of fact it has succeeded where other means have failed, and is worth a trial in many cases.

CYSTITIS.

Several cases of cystitis have recently been seen in the wards. In some an extension of gonorrhœal inflammation seemed to be the exciting cause; in another acute nephritis; in one persistent masturbation had given rise to an irritability of the bladder, which resembled cystitis in its symptoms. In all the cases there was frequent painful micturition, burning pain in the region of the bladder and the urine was found to be alkaline and to contain considerable mucus and some pus. The treatment pursued in most of the cases was the administration of salts of potash in the infusion of buchu, and the daily washing out of the bladder with a saturated solution of boracic acid. For some time past the favorite solution for injection into the bladder has been that of Thompson, viz.: ʒj of biborate of soda in ʒiv each of water and glycerine, but more recently the use of the saturated solution of boracic acid has proved of equal value. Its strength is about one part of acid to twenty-six of water. It is injected warm, two ounces being thrown into the bladder, allowed to remain about two minutes and then to flow out, and the injection repeated until all the mucus has been washed out and the bladder is left clean. In all cases alluded to, rapid improvement occurred in all the symptoms and final recovery.—*Bellevue Hospital.*

SORE THROAT OF TYPHOID FEVER.

Dr. DaCosta places his patients upon the use of the tincture of iron, giving fifteen drops every three hours. He thinks this is an excellent mouth wash for such cases:

 ℞ Alum;
  Borax, ... aa ℥ss ;
  Water, ... f ℥iii ss;
  Glycerine, ... f ℥ss. M.
 Signe.—To be used freely as a mouth wash.

If this solution is too strong it must be diluted. At the same time that it is being used, a light poultice with a little mustard in it is kept round the patient's neck.

TYPHOID FEVER.

The Bellevue Hospital treatment in all cases is aimed at the reduction of the temperature, the nourishment of the patient, and at supporting the action of the heart when signs of asthenia manifest themselves. It is desirable to keep the temperature below 103°. Whenever the mercury rises above this point antipyretic treatment is employed. Cold is used in all cases by wrapping the patient in a sheet which is kept wet with water at a temperature of 70°. This may be continued an hour if necessary, and may be repeated at intervals of four hours. It is found to be more effectual than the cold bath, and can be applied with less disturbance to the patient, as a rubber sheet can be placed beneath him in his bed, and the wet sheet at once used, without any transfer to a water bed or cot. The temperature usually falls 2° during the application of the cold, and continues to fall for half an hour after it is discontinued. The temperature is to be watched during the application, and the pack should be discontinued when it reaches 101½°. It is never to be continued until the thermometer indicates 99° lest the patient pass into a state of collapse. Cold may also be applied by sponging, and when the temperature is not very high this is often resorted to, each extremity being sponged separately, so as to expose the patient as little as possible. These two methods have superseded the use of the cold bath, as they are more convenient, more easily applied, disturb the patient less and give as good results. In severe cases the antipyretic action of quinine is added to that of the cold pack, a hypodermic injection of 15 grains being given at the beginning of the cold pack. In the third week of the fever some prefer to abandon the pack and trust wholly to quinine in large doses, fearing collapse or pulmonary complications.

For the nourishment of patients milk is used exclusively, and as much is allowed as the patient can take. Usually a full glass of milk is taken every hour, lime-water being added in the few cases where the milk alone seems to disagree. The proposal of Sir Wm. Jenner to substitute beef tea for milk, does not find much favor in hospital practice, but among private patients where beef tea can be properly prepared it is often employed. The demand is now so great that two of the New York restaurants prepare it daily for the use of the sick, and are able to dispose of large quantities.

In the latter weeks of the disease in all cases, and during its whole course in many asthenic patients, an important indication is to sustain the heart power. This is done by a free use of alcohol, whisky being the form most used at present. The quantity is determined by the needs of the case, but where there is necessity it is not spared, one case being allowed twenty ounces daily for three days with good result. Some use ammonia as well as whisky, but its action is considered less certain. When it is used the liq. ammon. acet. and the spts. ammon. aromat. are combined. Other symptoms are met by appropriate measures as they arise.

THE DIARRHŒA OF TYPHOID FEVER.

Prof. Roberts Bartholow recommends Tr. iodine, gtt.v., well diluted with water, in the treatment of typhoid fever Under this—one of the German so-called specific plans of treatment—he says, "with proper diet and nursing, the mortality is much diminished." The same writer, for the diarrhœa of that disease, prefers the following:

℞ Liq. pot. arsen., ... gtt. ii ;
   Tr. opii., ... gr. iv.
   Repeat as often as required. .

PELVIC EFFUSION.

The following are Dr. Maury's conclusions as to the treatment of pelvic effusions: 1. Caution and judgment are eminently demanded in the treatment of pelvic effusions; in the management of pelvic abscesses we should wait until maturation is complete, and simply assist Nature by making an incision as early as we are satisfied she has clearly indicated the point of opening. This is demanded in order to lessen the risk of a rupture into the peritoneum or bowel. 2. Inasmuch as many pelvic abscesses do not point at all, and manifest no tendency to open of their own accord, surgical means must be employed to make a way for their evacuation. 3. Generally these abscesses can be reached through the vagina, and whenever the effusion presents at the vaginal roof, so that it may be felt as a resisting body (it is not necessary that it should come down into the pelvis), it may be evacuated by the trocar. In rare cases these tumors present only in the rectum, or through the abdominal walls, and cannot be reached through the vagina. 4. Whenever we are satisfied of the existence of pus, and that ripening of the abscess has occurred, and thinning of the wall can be discovered, let us open it at once. 5. When we cannot, by physical signs alone prove the presence of pus, as is often the case, but believe it to be present from the constitutional symptoms, we should not hesitate to explore the pelvic roof or rectal or abdominal wall by aspiration, and remove the effusion without delay, whenever found. 6. The great majority of serous effusions will disappear under the influence of rest and counter-irritation. The very few which continue, in spite of medical measures, should be treated like similar effusions into the pleura. 7. Should such an effusion remain unabsorbed for three or four weeks after the beginning of the attack, and all acute symptoms have subsided, and especially if pain and a feverish condition be present, we should not hesitate to aspirate with a delicate trocar, and remove the effusion. 8. We are often unable to tell from the patient's history how long the effusion has been present, especially if the case has been sub-acute or chronic from the beginning: but we may always with propriety aspirate, if the condition is not one of acute inflammation, and if we are satisfied of the inutility of remedies.

CHOLERA MORBUS.

In cholera morbus Dr. Bartholow recommends the following:

℞ Chloral hyd., ... ℨ iij;
   Morph. sulph., ... gr. iv ;
   Aqua laurocerasi, ... f ℨj. M.
Sig.—From fifteen to thirty minims hypodermically.

This injection produces considerable burning pain, and sometimes an indurated lump, but is rarely, if ever, followed by suppuration.

FLAGGING HEART.

Dr. DaCosta calls attention to the use of digitalis hypodermically for the purpose of sustaining a flagging heart. Two drops of the fluid extract are equivalent in strength to fifty minims of the tincture. This amount (gtt. ii), well diluted with water, is what is generally used, and he has always found that it answers all the purpose of hypodermic medication excellently. This dose can, of course, by repeated as often as necessary.

MALARIAL FEVER.

Dr. John Shrady, of New York, employs with success the following method in the case of children where great irritability of the stomach exists, and where the internal administration of quinine is otherwise contra-indicated:—

Denude the cuticle at the epigastrium by means of a blister, of the size of 'a half a dollar, and then apply the following:—

℞ Sulph. quiniæ, ... ʒj;
   Cerat. simplicis, ... ʒj. M.
Sig.—Spread on lint and renew the dressing morning and night.

He has in this way obtained the characteristic effects of the salt, as indicated by the decline in temperature and the abbreviation of the paroxysms of fever. As the surface of the sore ceases to be absorbing after the third day, a repetition of the process in the immediate neighborhood may be, although rarely, necessary.

FETID CORYZA.

In chronic coryza, with fetid secretion—a most unpleasant affliction—Dr. J. Solis Cohen says that the most efficacious wash he has employed is

℞ Aquæ chlorinii, ... ʒj;
   Aquæ, ... Oj.
Use with the nasal douche.

The skin, kidneys and bowels must be kept active by frequent bathing and an occasional purge. A cure requires months of treatment.

LINEA SYCOSIS.

Dr. L. A. Duhring uses:

℞ Hydrarg. sulph. flav., ... gr. xv;
   Adipis, ... ʒj. M.

SEBORRHŒA CORPORIS.

The same authority employs:

℞ Sulphur. precip., ... ℨss:
    Adipis, ... ℨiv. M.

This is to be well rubbed into the affected parts, morning and evening.

DERMATITIS.

This Dr. Duhring's formula:

℞ Ext. grindeliæ rob., ... ℨij;
    Aquæ., ... Oj. M.

TREATMENT OF ANASARCA.

In the treatment of general anasarca in Bright's disease, the necessity of exciting the skin to action is universally recognized. By means of the perspiration, much fluid can be removed from the body and the œdematous condition of the patient relieved. In many cases where no danger exists of the occurrence of œdema of the lungs, pilocarpine may be used to produce this effect. In some cases it is contraindicated by organic or valvular disease of the heart. In these cases the following method is employed in the hospital: The patient is sponged off with alcohol, and is then wrapped in a wet sheet, over which several blankets are placed. In the course of an hour the diaphoresis is usually profuse. The use of alcohol before the pack is recent, and has proven successful. It is supposed to act directly upon the sweat glands as a stimulant, and certainly increases the amount of the sweating produced by the pack. When these means do not produce sufficient diaphoresis, the fluid extract of jaborandi in doses of one drachm is given just before the pack is applied, and as it may produce nausea if given by the stomach, a preferable method of administration is by enema, in which case the dose may be increased to one and one-half drachms. In one case, at present, in the wards this method is daily pursued with good results. The anasarca is rapidly decreasing. When diaphoresis by the hot air bath is attempted, the use of alcohol is found to be of equal service, and in cases of uræmic convulsions it has certainly hastened the excretion through the skin.—*Bellevue Hospital.*

THE PNEUMONIA OF TYPHOID FEVER.

Dr. DaCosta's treatment is by turpentine internally, with morphia enough to keep the turpentine from irritating the bowels and bladder. His local treatment of the chest consists in the use of dry cups and turpentine stupes. Twelve grains of quinia are given daily. Dr. DaCosta is a strong believer in the value of turpentine in the pulmonary complications of typhoid fever. In the vast majority of such cases stimulus is needed, about eight ounces of whiskey and four ounces of wine being given in the twenty-four hours.

SCIATICA.

Ten cases of sciatica occurring in the Pennsylvania Hospital were treated by Dr. DaCosta with deep daily hypodermic injections of chloroform, near the seat of pain. After a few days, if improvement was not marked, a mixture of ten grains of the iodide of potassium with two drachms of ammoniated tincture of guaiacum were ordered.

RHEUMATIC ENDOCARDITIS.

Dr. Alonzo Clark says that there are really only two things of much importance in this connection, viz.: (1) abstracting blood and (2) quieting the heart with opium or some other sedative.

INTESTINAL HÆMORRHAGE IN TYPHOID FEVER.

Ergot by the mouth with hypodermic injections of whisky, with a little ammonia, constitute Dr. DaCosta's main reliance. The dose of ergot is twenty minims every three hours. The injection is composed of from 3 to 5 drops of aqua ammonia with 30 minims of whisky.

THE DELIRIUM OF TYPHOID FEVER.

Dr. DaCosta gives a one grain suppository of opium three times a day. It quiets the bowels, if there is any tendency to diarrhœa, and has a good influence over the disturbed action of the nerves. When the opium does not act speedily by the bowels, he gives one-sixth of a grain hypodermically. He finds the combination of digitalis and opium a good one—best suited to cases where there is also rapidity of the heart's action. To calm the nervous system the bromides are given in twenty grain doses. When this fails, fifteen grain doses of chloral usually are effective. In these cases cold applications to the head are also indicated. Putting the patient in a warm bath will generally be found to relieve the delirium and wakefulness, and produce sleep.

PNEUMONIA.

The success of the treatment of the high temperature of typhoid fever by the cold pack, has led to its employment in a number of cases of pneumonia attended by great febrile movement. The fact is recognized that it is only in the minority of cases that such treatment can be used But when a patient has been, previous to the invasion of the disease, perfectly healthy, when the action of the heart is strong, and the general condition of the patient favorable, the cold pack is regarded as of great service.

One case was recently seen in which the rational symptoms and physical signs allowed no hesitation in this diagnosis. The man was admitted to the hospital three days after the initial chill, with well marked signs of solidification of the lower lobe of the left lung. The next morning his temperature was 104° and at noon 105°. He was then placed upon a water bed and a wet sheet wrapped about him, which was sprinkled with cold water every fifteen minutes. In the course of two hours the temperature had fallen to 103.5°, when the pack was removed.

In German hospitals the antipyretic treatment of pneumonia is succeeding. It does not seem improbable that it is soon to meet with equal favor in this country. The course of the disease cannot be affected by it, but the most exhausting of its symptoms, viz.: the high temperature, can be controlled, and the chances of recovery largely increased. The use of quinine with the application of the cold pack doubtless makes the effect of the latter in depressing the temperature more lasting. But even where it is employed alone the result is more satisfactory.

In addition to antipyretic treatment, patients who are suffering from pneumonia are given as much milk as they can digest; a half ounce of whiskey every two hours, where a heart stimulant is indicated; and as an expectorant mixture, ammonium carbonate, five grains with spirits chloroform, fifteen drops, every two hours. The last is of great service in lessening the tenacity of the muco-purulent sputa, and in allaying the irritation of the air passages.
—*Bellevue Hospital.*

SEROUS DIARRHŒA OF INFANTS.

Dr. Hutchins, of Brooklyn, has treated twenty-seven cases of this disease in infants between the ages of two months and two and a half years, within the past three months, with this drug alone, and in all the cases the disease was promptly and permanently controlled. The dose was from 3 to 5 grains every two or four hours. The vomiting was controlled as soon as the medicine began to show its effects on the discharges, and the drug was well tolerated.

GASTRIC VERTIGO.

Dr. Bartholow has seen the greatest improvement in these cases from Fowler's solution, two drops being given just before each meal; it is of the greatest service in stomachic vertigo.

THE TREATMENT OF ANEURISM BY ELECTROLYSIS.

According to Dr. Pepper the mode of operating is very simple. You need two sharp platinum needles, coated with gutta-percha, and after freezing the skin with ice or with ether spray, you should plunge one needle, previously connected with the galvanic battery, boldly in with a single stroke, until you feel all resistance cease. The second needle is to be introduced in the same way. Thus far it will cause but little pain, but the moment the current is turned on the heart will give a great bound and the pulse become greatly accelerated. This should not, however, be any cause for alarm. Gradually turn on the full current and leave it on for some minutes, when the operation is completed by withdrawing your needles. Electrolysis is only applicable when the medical treatment has been tried without success, and when the aneurism comes up close to the thoracic walls; but this is so often the case, that it can be frequently applied.

FLOATING KIDNEYS.

Medicines directly can do very little, but indirectly they can accomplish something in the way of relieving symptoms. An abdominal bandage well applied gives great comfort. In order to adjust it the patient is placed recumbent, and then the offending kidney is gently pressed into position, where it is held by the bandage, drawn with considerable firmness, and it should be worn night and day; if taken off at night it will undo the work of the day. By constant use of the bandage for months and years adhesions that will maintain the kidney in place may occur, and Dr. Bartholow has thus succeeded in obtaining new attachments when the bandage was worn for several years. Attention to the functions of the abdominal organs is also essential; flatus must be carefully guarded against by cutting off all articles of food that ferment readily in the stomach or intestines.

He gives also, with a view of overcoming this condition of abdominal fullness, a drop or two of carbolic acid three times daily. He also prevents constipation by aperients occasionally, so that no colonic accumulation may press upon the kidney.

TAPE WORM.

This is Dr. Tausky's treatment. Three ounces of the pomegranate root is soaked for twenty-four hours in eight ounces of water. This is then boiled down to three ounces, to which is added three ounces of the etherial extract of male fern, one and one-half drachms of sulphuric ether, two drachms of fluid extract of valerian, one drop of croton oil, and one and one-half ounces of honey. The patient then abstains from all food, save herrings and onions, and also from water.

Dose: One-third of the above every half hour. As soon as the patient feels intestinal contractions or colic, whether this was after the first, the second, or third dose, one ounce of castor oil is administered hourly, in cold, black coffee, without sugar, until the worm is expelled. In case the vermicide mixture causes nausea, lemonade, ice pills, or strong, cold, black coffee without sugar, relieves it. The preparatory treatment is a dose of castor oil forty-eight hours before taking the worm-medicine, so as to empty the bowels thoroughly of fæces. Milk diet must be used exclusively for the twenty-four hours following the laxative, and during the twenty-four hours before taking the mixture the patient must abstain from food and drink, except that he occasionally may take a little salad, made up of salt herring, onions, and garlic. After the worm is expelled, mucilaginous food only is taken for a day or two, and a starch and laudanum enema is administered in case of tenesmus.

ANÆMIA OF CHILDREN.

Dr. Jacobi thinks that barley-water and cow's milk make better muscle than poor, mother's milk. He avoids solid food, in the main, for infants; and avoids cow's milk, either undiluted or diluted with water only. No milk is used without the addition of some gelatinous or farinaceous decoction, such as barley-water, etc. In anæmia beef soup is added. Solid food may be given at the end of the first year, and such articles slowly added to the diet list as physiology and experience permit. Irregular and fast eating are prohibited; out-door exercise is enforced; crowded school-rooms and excessive private lessons are avoided.

Among the remedial agents iron has long been resorted to in the treatment of anæmia and chlorosis by him. Whether or not it is the iron which produces the beneficial effect has not been answered to the satisfaction of all; for, a great many of that class of patients recover in consequence of change of diet, with rest and an improved general nutrition, and without the use of any iron whatever. Besides, there are a number of cases in which the administration of iron has been absolutely unavailing. Moreover, there is plenty of iron in almost every article of food. Certainly the doses usually given are large when compared with the iron contained in the food, and with the amount of iron (gr. xlviss and no more) present in the whole quantity of blood circulating in the human body.. But it has not been found whether the iron does not act in some way other than by increasing the amount of the metal contained in the hæmoglobin. There is no doubt, according to Dietl and Heidler, but that it is absorbed in the stomach, and also, very probably, in the upper part of the small intestine. It reappears in the bile, the pancreatic juice, and the intestinal secretions, not only after it has been taken into the stomach but also after it has been injected into the veins. The preparations which, in his opinion, are the most beneficial in the anæmia of children are the lactate, the tincture of the pomate, the pyrophosphate, the subcarbonate and the tincture of the chloride. The syrup of the iodide is indicated where, in addition, an absorbent is required; as for example in slow convalescence after inflammations resulting in exudation, and specially in disease of the glands and lungs. The subcarbonate combined with three times by weight of subcarbonate of bismuth, and three or four times by weight of bicarbonate of soda, he considers as especially beneficial when gastric catarrh interferes with general improvement during slow convalescence or progressive anæmia. The tincture of the chloride must be regarded as a vascular irritant, and, whenever the action of the heart is lowered and blood pressure is lacking, is *the* preparation which will be found to be the most beneficial.

It has seemed to him that the pyrophosphate is the preferable preparation in cases of anæmia with gastric catarrh, and catarrh or digestive incompetency of the upper portion of the small intestine. The compound hypophosphites and the phosphates he has used with good results, notwithstanding the fact that their elimination is nearly as rapid as their ingestion.

In cases of chronic anæmia he has also used arsenic with benefit in minute doses daily after meals, and well diluted with water, especially in a peculiar torpid condition of the stomach which will not digest and assimilate in consequence of absence of both nerve power and gastric juice. It may be given with iron, with or without stomachics. Strychnia, also, in his hands, has proved very useful as an adjuvant to either iron or arsenic, and to a child two years old it may be given with safety in doses of one-fortieth of a grain daily, and continued for a long time. He has also used phosphorous in substance, in minute doses, with good results.

Cod-liver oil, in many cases, is beneficial, but frequently the contraindications are overlooked. Most children do not bear it well in the summer time; some do not bear it at all. At all events, it should not be forgotten that, whenever digestion is impaired and gastric catarrh is present, preliminary treatment is required before the administration of either cod-liver oil or iron.

SCARLET FEVER.

If there is a membranous production in the throat Dr. Alonzo Clark uses lime-water from a spray producer. He has the child's mouth opened and throws the spray into the mouth when the child is taking an inspiration—showering the fauces with the lime water. The doctor believes that this lime-water breaks up the attachments of the membrane to the living tissues. He is a believer in the use of the wet pack.

When the eruption is of a black character—i. e., when there is hæmorrhage under the cuticle—he gives quinia and the vegetable acids freely.

When œdema occurs he advises the immediate employment of a warm bath, and that the child be kept in bed in a warm room after being taken out of the bath, and that enough clothes be kept over it to maintain a constant, gentle perspiration. The bowels must be kept free and the food unirritating.

CONVULSIONS IN CHILDREN.

Dr. A. A. Smith thinks that, whatever their cause, convulsions should be arrested by the use of chloroform. Convulsions dependent upon the pain of teething he controls first by opium and then by the use of the gum lancet. Convulsions due to intestinal worms he treats by a combined cathartic and opiate. Convulsions due to a malarial poison he finds yield more promptly to opium than to anything else. When the paroxysm has been controlled by the opium, the child is put thoroughly under the influence of quinia. For a child over four years of age, Dr. Smith advises the use of paregoric, but if the child is under four months he uses, every half hour, a teaspoonful of a mixture of bromide and chloral, with bi-carbonate of sodium, of which mixture each dose contains one grain of each remedy. He rejects the use of hot baths in convulsions, as likely to add to the excitement of an already disturbed nervous system. With the object of keeping the nervous system free from agitation, he forbids the hot bath, and gives orders that the room be kept perfectly quiet and darkened, that plenty of fresh air be admitted, and that the opening and shutting of doors be avoided.

To control convulsions dependent upon high temperature he recommends veratrum viride, combined with opium, two drops of the tincture being given every hour to a child from six to eighteen months of age. He thinks favorably of the sedative action of calomel.

FATTY HEART.

Dr. DaCosta finds that these cases do best upon stimulus. He prescribes f ʒss. of whisky to be taken at each meal. Digitalis is not advisable, and is only resorted to to control the action of the heart. As a tonic to the muscles of the heart he orders the $\frac{1}{30}$ of a grain of strychnia thrice daily. To control the attendant dyspnœa he has had the best results from the application of dry cups. When it occurs in paroxysms he gives the aromatic spirits of ammonia internally.

GOUT.

Dr. Janeway's treatment consists in giving colchicum, the wine of the seed or root, which ever you prefer. He generally uses the wine of the seed; dose, half a teaspoonful. If it produce nausea and vomiting, he decreases the dose. He gives with it Vichy water, or the alkalies. For the effusion he applies iodine over the joint. He thinks, too, that bandaging does good, especially where there is swelling and thickening of the tissues. Then after having followed out this course a week or so he gives his patient the iodide of potassium.

ECZEMA.

The Moist Form.—Dr. George Henry Fox promotes the surface exudation, and relieves the swelling and itching in great measure by applying thin, sheet rubber. Then, for the purpose of soothing and protecting the thin, tender, newly-formed epidermis, and of preventing the leg from swelling again he applies night and morning the ordinary zinc ozide ointment on strips of cloth, with a muslin bandage.

The Dry, Erythematous Form.—Local treatment is uncertain and transitory in its effects. The patient's general health is improved by tonics and alkaline diuretics.

Eczema Squamosum.—Acetate of potassium is given internally and the ointment of Cade applied locally.

Chronic Eczema of Upper Lip.—The first necessary step is to improve the patient's general health. The lip must be shaven, or the mustache trimmed as closely as possible. The ammoniated mercury ointmeut is rubbed gently upon the lip night and morning. Where the eczema is the result of a thin, serous, acrid nasal discharge the treatment is directed to the removal of the cause.

ACNE.

In strong and otherwise healthy subjects, Dr. Fox recommends careful diet and local frictions with his prepared olive soap.

Where the eruption is due to indigestion antidyspeptic treatment is pursued together with soap-frictions.

In the highly inflammatory form occurring in ill-nourished and debilitated subjects the general condition of the patient is to be looked after.

The local treatment is limited to the most soothing methods. The face is bathed frequently with very hot water and a lotion of borax and glycerine in rose water or a little cold cream is applied.

PEDICULOSIS CAPITIS.

The simplest and most effective form of treatment in Dr Fox's estimation consists in the free use of kerosene oil. He does not think it necessary to crop the hair.

OBSTINATE CONSTIPATION.

Every night before retiring Dr. Da Costa orders a tablespoonful of sweet oil to be taken and thrice daily after meals he gives gtt. j of the fluid extract of belladonna in f ℥ i of the compound tincture of cinchona.

DILITATION OF THE STOMACH.

Dr. Da Costa thoroughly washes out the stomach, first

every second day and then twice a week. This washing is done with the stomach pump tube with piston and syringe attached. The water used is rendered slightly alkaline by the addition of a small amount of the bi-carbonate of sodium.

To counteract fermentation at first gr. j. of carbolic acid are given thrice daily in water and later f ℥j. of sulphurous acid is given thrice daily, largely diluted with water.

During convalescence the washing is discontinued and gtt. x of the tincture of nux vomica substituted in place of the sulphurous acid.

OBSTRUCTIVE JAUNDICE.

Dr. Janeway's treatment consists in subduing the catarrhal inflammation in the stomach and intestine by counter irritation, by sedative applications, by light diet; and in the second place, if there be any uncomfortable symptoms from the jaundice, in removing them. For instance, these patients often complain a good deal of the fetor of the evacuations; they complain more or less of wind, and colicky pains which are due to a failure of entrance of the bile into the intestine. And then again they are very apt to complain of an itching of the skin. It is generally recommended in the treatment to give alkalies, and sometimes to give an emetic. If the obstruction is due merely to mucus he gives an emetic. He once tried to produce removal of the obstruction by giving an emetic in the form of apo-morphia. It was when it was first coming into use as an emetic. It produced great prostration, and he does not advise its use in a case of jaundice. He has not done so since. The patient came very near having syncope, and the next day a severe bleeding of the nose came on, which was very difficult to control. It so happens that in some of these cases of jaundice hemorrhage comes on very easily, and is difficult to control.

ACUTE RHEUMATISM.

1. In the feeble, anæmic, nervous subject, Dr. Bartholow gives tinct. ferri chlorid., ℳ xxx every four hours; orders the joints to be kept at rest, wrapped in cotton if the patient desires it; and if they are very painful, small blisters (the size of a silver dollar), to be applied around them. An occasional laxative of Rochelle salt is added. He thinks that the iron cuts short the disease, lessens the danger of cardiac complication, and also has the power, as Anstie pointed out, of preventing impending attacks. The blisters relieve pain, and bring about a more alkaline condition of the blood and urine. Thus treated, his cases of this type rarely last more than two weeks, heart complication is infrequent, convalescence is rapid and relapses uncommon.

2. Fat and flabby subjects require the alkaline plan: Two drachms of potassium carbonate, ½ drachm of citric acid, and four ounces of water, are given every three or four hours, until the urine ceases to be acid, when the amount is reduced one half, the reduction being then continued daily until the fourth or fifth day, when, if the urine continue alkaline, quinia (six grains every four hours), or preferably tinct. ferri. is added. If the attack is severe, blisters are applied. With this treatment this class gets well within two weeks.

3. Vigorous subjects, often with hereditary tendency, are often promptly relieved by salicylic acid in scruple doses. Not less than ʒij, is administered in twenty-four hours, and considerably more may be required. It has been found to be more effective given in solution with an excess of alkali. A cure is thus not unfrequently effected in three or four days, but some stomachs cannot bear it, and if it depress the heart it is stopped. If after three or four days it produce no improvement, it is useless to persist in it. In all forms the diet used is liquid. Opium is thought objectionable because it checks elimination; atropia promotes elimination, and is therefore preferred as an anodyne,

being given hypodermically in the neighborhood of the affected joints, and it is rarely necessary to exceed gr. $\frac{1}{80}$ a day.

Should cardiac complication arise, the carbonate of ammonia (gr. v, doses frequently), and infusion of digitalis, with hypodermic injection of morphia are given at once, to dissolve fibrin, check inflammation, and lessen the work of the heart. When the acute symptoms have subsided, iron and quinia are substituted for the ammonia and morphia. Experience has also shown a blister on or near the præcordia to be useful.

In the sudden hyperpyrexia (fortunately very rare), where the temperature leaps without cause to 106°—109° F., the cold bath has been found to be necessary to ward off certain death.

PAROTID SWELLING FOLLOWING TYPHOID FEVER.

Ice is steadily applied to the swollen gland. If suppuration takes place in spite of the ice, poultices are at once applied, and the pus released by an early incision. Dr. DaCosta's internal treatment consists in gtt. xx of the tincture of the chloride of iron every three hours, and gr. xii of quinia, and f℥ iv of whisky daily.

GRANULAR LIDS.

Dr. Cornelius Agnew uses:

R Acid. tan., ... gr. v;
Glycerinae, ... ℨij;
Sod. bibor., ... ℈ij;
Aq. camph., ... f℥j. M.

Sig.—To be used in the spray, care being taken not to allow the spray to fall upon the cornea, unless it is vascular or cloudy.

CHRONIC DIARRHŒA AND PERITONITIS.

Dr. DaCosta injects ♏v. of Magendie's solution into the abdominal walls every morning and evening. This treatment has been always followed by most gratifying results.

EXOPHTHALMIC GOITRE.

Dr. DaCosta gives iron in anæmic cases. He has derived the most decided advantage from the administration of a course of digitalis and belladonna. Where there is a large amount of cardiac enlargement he uses veratrum viride, and thinks it better than aconite—dose gtt. i of fluid extract thrice daily. Rest in bed is regarded as a very important item of treatment. The thyroid enlargement is reduced by the continuous local application of ice.

CHRONIC NASAL CATARRH.

Dr. Harrison Allen applies the following locally:

℞ Acid. carb. pur., ... gr. v;
Fl. ext. geran. mac., ... gtt. xv;
Glycer. distillat., ... gtt. x;
Pulv. iodoform., ... ʒiij ss;
French gelatine, ... ʒi;
Aquæ, ... q. s.

Dissolve the gelatine in a little water, then add the other ingredients, and rub to a smooth paste.

This prescription may be made up without the geranium.

PLEURISY AND EMPYEMA IN CHILDREN.

Dr. J. Lewis Smith regards blood-letting, and that by leeches alone, adaptable only to primary pleurisy. After leeching he is in the habit of applying a poultice composed of one part of mustard, and sixteen parts of flaxseed. This poultice is covered with oil silk and changed every twelve hours. He regards blistering as inadmisable in the early stages. In the first stage he uses ancoite and quinia, combined with some opiate. He considers digitalis as particularly useful where the pleurisy is complicated by pericarditis. His favorite opiate consists of Dover's powder, tincture of the root of aconite, and balsam of tolu. When pus exists in the pleural sac he thinks its removal by operative measures indicated. If the fluid is sero-fibrinous he does not operate. He considers watery cathartics as depressing. As diuretics he employs the acetate and iodide of potassium. When he aspirates he uses a trocar and aspirator combined, and does not employ a canula with a sharp point for fear of hurting the lung. He regards a puncture ⅜ of an inch deep sufficient. When the trocar is detached, the aspirator is attached to the canula and the fluid emitted. A single operation he regards as enough, when the fluid is sero-fibrinous. He establishes a fistulous opening in empyema. His method is to remove the liquid, wash out the pleural cavity with carbolized water (temperature 100° Fhr., strength one part to one hundred,) and then introduce threads by means of an instrument which he has devised for the purpose. These threads are allowed to remain. Unless the pus is fœtid the washing out is to be done *but once;* if fœtid every second day.

OBSTINATE MALARIAL ATTACKS.

After checking the disease with quinia, the paroxysm will recur, and the treatment will thus often be brought into discredit, unless some few points are borne in mind as regards the method of administration. Give the quinia at least three hours before the expected paroxysm. Shall we give small doses frequently repeated, or large doses less often? The latter is the true mode. You will then give fifteen grains three hours before the expected paroxysm. I prefer this to the former method, for this reason, which I regard as indisputable. Quinia, though not eliminated from the system with great rapidity, yet is eliminated, and chiefly by the urine. If we were to give it in small doses early in the morning, by afternoon it would be eliminated; and would require to be repeated, and in larger amount, in order to check the paroxysm. Therefore, it is more economical, as well as more effective, to give a single large dose, which is also more agreeable to the patient; for I affirm that fifteen grains given at once will give much less distress than one grain every hour until the same amount be taken. Large doses obtund the sensibility of the cerebral centres, while smaller ones cause excitement of the brain and tinnitus.

By giving a single large dose of fifteen grains of quinia at least four or five hours before the time for the appearance of the expected chill we break up the paroxysms. What shall we do to prevent their return? We ordinarily hear that the chills are apt to return at septenary periods; but if you will look into the matter you will find that they recur in multiples of the original number. Thus, tertian would return in six days, or if not, then on the ninth, twelfth, fifteenth, eighteenth, or the twenty-first day; and in quotidian they are apt to be manifested in multiples of two. On these critical days, the remedy should be repeated. If we break up the chill to-day, on the day after to-morrow, although he may not have a decided chill, he will have some significant symptoms, that are evidences of systematic disturbances; he will excrete more urine, he may have a diarrhœa, general muscular soreness, or something else indicating the influence of the malarial poison. We must, therefore give our quinia again, and repeat it on subsequent days, multiples of the original attack, administered in anticipation of the former hour of the attack. On the morning of the sixth and seventh, the thirteenth and fourteenth, the nineteenth and twentieth and twenty-first days, doses of ten grains should be given.

What else? Do you abandon your patient in the interim? Ten grains of quinia will not be sufficient to relieve a damaged liver, or to reduce an enlarged spleen; in other words, the condition of chronic malarial poisoning. Treatment must be directed to this object as well as to breaking up the chills, or they will inevitably return. Lugol's solution, in five-drop doses, given in water before meals, and Fowler's solution, three drops after meals, always prove most efficient aids. It is best, about the twenty-first day, to give a full antiperiodic dose of quinia for three days, for by this time there is a much greater accumulation of morbid material in the blood than at the other periods named.

Please bear in mind these rules which I have just given you, for you will find that they will stand you in good stead in all these cases of obstinate malarial attacks.—*Roberts Bartholow.*

LUMBAGO.

According to Dr. Alfred Stillè, the best treatment in acute lumbago, at first, is the application of wet-cups to the muscle or muscles affected, to be followed immediately by narcotic fomentations in the shape of a bag of hops soaked in hot water, hot vinegar, or alcohol, applied directly over the scarified parts. There are various stimulating and anodyne liniments which he also uses such as turpentine, ammonia, and camphor. Opium in the form of a ten-grain Dover's powder, given early, relieves pain and produces diaphoresis. Atropia hypodermically (one eightieth of a grain) is valuable, but must not be given to nursing women. Morphia may also be given hypodermically (except in pregnancy) and these two remedies are usually the best in private practice when wet-cups cannot be nsed. Iodide of potassium, in doses of five to ten grains every three hours, gives very good results. The most useful class of remedies in chronic lumbago are blisters, sinapisms, the actual cautery, etc. Local friction and *massage* conscientiously applied are often useful when counter-irritants fail. Tepid water may be applied, either in the shape of wet compresses kept in constant contact with the part, or in the form of a douche falling steadily upon the rheumatic muscles for some time from a height of eight to ten feet. The action of water, though slow, is a very permanent one. After the treatment by douche or by wet compresses the parts should be briskly rubbed with a coarse cloth or a skin-brush, and then covered with cotton or wool or a piece of India-rubber cloth. The use of a metallic brush is sometimes advantage ous, and finally tying the cloth over the lumbar regions and ironing them thoroughly two or three times every day, following this up with the application of some stimulating liniment, is often to be advised.

CATARRH OF THE STOMACH.

Acute Catarrh.—The most essential point in the treatment is a proper regulation of the diet. The most uniformly applicable food is milk. Care must always be taken to secure milk of good quality.

When there is considerable nausea and frequent vomiting, from two to three fluid ounces of milk, with one fluid ounce of lime-water, every two hours, will usually be retained without difficulty. Such a plan, supposing the feeding to be discontinued through the night (which, by the way, should always be done, unless otherwise demanded by excessive prostration, if there is a tendency to sleep,) supplies about sixteen or twenty-four fluid ounces of milk per diem. If the irritability is too great for the retention of these doses, the same combination of milk and lime-water must be given in diminishing amounts at correspondingly short intervals, until the proper measure, if it be only a teaspoonful, is reached. On the contrary, when the vomiting is less obstinate, or as the patient improves, the doses and intervals should be increased, and as opportunity offers, the diet gradually extended, by the addition of farinaceous articles, broths, chicken, etc., until the ordinary food is resumed. Very exceptionally, milk and lime-water cannot be retained. When this happens, it is best to abandon the milk diet altogether and substitute carefully-prepared beef-tea or chicken or mutton-broth, entirely free from fat, in doses of two fluid ounces every three hours. Beef-juice is also serviceable under these circumstances. Dr. Starr has never had occasion to use either whey or artificially-digested milk, each of which is highly recommended in obstinate vomiting. Thirst, which is often a distressing symptom, is best relieved by the moderate use of ice, the ingestion of large draughts of water tending to prolong the vomiting. Until the active symptoms have subsided, confinement to bed is necessary.

Having regulated the diet and enforced complete bodily rest, the patient is put far on the way to recovery; but still much may be done by medication to shorten the illness. At the beginning of the attack, if the bowels are obstinately confined, and particularly if the skin or conjunctiva are at all yellow, he directs in the evening three grains of "blue mass" or five grains of calomel, to be followed in the morning by a Seidlitz powder; when the constipation is moderate, the lower bowel is simply evacuated by an enema. At the same time, either bicarbonate of sodium or citrate of potassium in the form of mistura potassii citratis, or preferably the "effervescing draught" is prescribed. The bicarbonate of sodium, employed most frequently, is administered in ten-grain doses, three or four times daily, mixed with a tablespoonful of milk or compressed into a pill. The citrate of potassium is usually reserved for cases where considerable fever is associated with the sick stomach. The "effervescing draught" is much more agreeable and efficient than the "neutral mixture;" it is ordered in two solutions—one composed of two drachms of citric acid to four fluid ounces of water, the other of one drachm of bicarbonate of potassium to three fluid ounces of water; half a fluid ounce of each are put together and taken during the effervescence, the dose being repeated every two or three hours. This mixture is yet more pleasant if equal quantities of lemon-juice and water are substituted for the nitric acid solution. In addition to these medicines, sinapisms or linseed poultices are applied to the epigastrium.

Chronic Catarrh.—A careful regulation of the diet is almost as important in the treatment of chronic as of acute catarrh of the stomach. Where there is much irritation, a diet of milk and lime-water is to be selected. The stomach is given an opportunity to rest and recuperate, by removing the cause of irritation and diminishing its work. No alcohol is to be allowed, and injurious occupations are modified as far as possible.

Very much may be done by medication. The treatment by alkalies is perhaps the most uniformly successful. The best alkali is bicarbonate of sodium; it may be given, when there is no decided irritation of the mucous membrane, with compound infusion of gentian or with infusion of colombo or quassia, either bitter adding to the efficiency of the soda. In very chronic cases, nitrate of silver may be prescribed with advantage; to produce good results it must be given when the stomach is empty. Dilute muriatic or nitro-muriatic acid, in combination with a bitter, may be used from the out set in atonic cases, but when there is an element of irritation they should not be employed until after a course of bicarbonate of sodium. For the habitual constipation, he has lately used, with very satisfactory results, in very obstinate cases a pill composed of ext. belladonnæ, gr. i; ext. colocynth. comp., gr. ij and ol. cari. gtt. ss., administered at bedtime. The painful sensations in the epigastrium are greatly relieved by counter-irritation.

CHRONIC HYPERTROPHY OF THE TONSILS.

Dr. Penrose's treatment of these cases is both general and local. He regards the general or constitutional treatment as very important, particularly as such patients are generally cachetic. Unless this general treatment is rigorously carried out, the local treatment is of no avail. The first thing to be done is to remove any bad hygienic surroundings which may exist. He then begins with mercurial purgatives and revulsives. He gives from one-twelfth to one-half of a grain of calomel every two hours, in a little sugar, until the child has from three to six movements every day. This calomel treatment is continued every two or three days for several weeks. At the same time, as a stimulant to the system, the child's whole body is thoroughly rubbed with hot whisky before going to bed

If the patient is decidedly weak, one fluid-drachm of brandy, well diluted, is given internally from three to six times daily. Patients have been kept upon these daily rations of stimulus for weeks at a time. Rubbing the body with hot whisky has been followed by the most excellent results. The other items of the constitutional treatment consist in the use of from three to six grains of quinia daily, and from five to ten drops of the tincture of the chloride of iron, from three to four times daily, in water. At the same time the child is fed up well with plenty of good soup, milk, eggs, and extract of malt. The object is to rebuild the whole fabric of the body, at the same time that all the diseased products are being removed in the alvine evacuations.

The local treatment is regarded as very important, but the local treatment is of no avail unless the constitution is regenerated by means of the remedies already indicated.

If the inflammation and hypertrophy are considerable, Dr. Penrose is in the habit of applying a solution of the nitrate of silver (sixty grains to the fluid ounce of distilled water) to the parts, by means of a brush. The application is made carefully and thoroughly. A small brush is used, and care is taken not to catch up too much of the astringent solution on the brush. The effort is made to secure the child's co-operation, and great care is taken that the brush is properly secured to its handle.

Powdered alum has been found to be another very excellent application in these cases. The following formula is ordered:

℞ Pulv. alum., ... ℥ij;
   Sacchar. . . . . q. s.
   M. et in chart. No. xx. div.
Sig.—One powder every two hours.

The child is taught to take a pinch of the powder every now and then, and not to swallow the whole thing at once —this with a view of prolonging the contact of the medicine with the diseased tonsils. When the child is large enough, lozenges of the chloride of potassium are often very efficient.

Another mild but excellent local application is the tincture of the chloride of iron with glycerine—one fluid ounce of each being applied by brush, once or twice a day. But very few cases resist this combined treatment. The constitutional treatment is kept up until the tonsils have been reduced to their proper size.

If the child is pigeon-breasted, a great deal has been accomplished by careful gymnastic training. Dr. Penrose is very much in the habit of recommending the use of three-wheeled velocipedes for the purpose of developing the chest, or, as a systematic plan of treatment, he has the child stand with its back against a wall, tells it to take a full breath, and then, when its lungs are thoroughly expanded, institutes firm pressure against the sternum.

ENDARTERITIS.

Dr. Bartholow has great confidence in the value of small doses of opium as a cardiac and nervous sedative. His patients in the Jefferson Medical College Hospital receive five drops of the deodorized tincture of opium every four hours. To arrest the usually accompanying condition of chronic arteritis, the hypophosphites, cod-liver oil, and quinine are employed. One fluid drachm of the lactophosphate of lime and three minims of Fowler's solution are given thrice daily, in addition to the opium. When any improvement is visible in the condition of the patient, quinia is given in energetic doses. It has been shown to have a good effect upon the coats of the arterioles.

MULTIPLE SCLEROSIS.

According to Dr. Bartholow, the therapeutic indications in these cases are the carbonate and the iodide of potassium. Five grains of each of these salts are ordered three times a day, before meals. From half a fluid drachm all the way to one fluid drachm of the syrup of the iodide of iron is given after meals.

ASTHMA.

Dr. Bartholow has succeeded in affording great relief to sufferers from this distressing complaint by the administration of fifteen grains of the iodide of potassium and twenty grains of the bromide of potassium four times a day. This combination has been found to be particularly useful where there is any spasm of the bronchi.

SCLEROSIS OF THE LIVER.

The watery purgatives—the so-called hydragogue cathartics—are the chief means employed of getting rid of the ascites. Dr. Bartholow orders from one or two drachms of the compound jalap powder every morning. With a view of acting upon the liver, and stopping the sclerosis he orders the one-twentieth of a grain of the chloride of gold and of sodium thrice daily in pill form. Chloride of gold acts on the hepatic cells and retards the hyperplasia of connective tissue.

DIPHTHERIA.

The two main indications, in the opinion of Dr. J. Solis Cohen, consist (1) in keeping up a supply of nourishment and stimulus and (2) in providing for the detachment and discharge of the morbid accumulations, when they threaten to occlude the air passages. The sick room must be systematically disinfected. This is done by the free use of sprays of carbolic or sulphuric acid. Solutions of the sulphate of iron or some other disinfectant are kept in all the vessels which are brought into the sick room to receive the discharges, the soiled clothing, refuse food, and slops of the patient.

He regards the chlorine compounds as of more efficacy in diphtheria than all other remedies. Of these he prefers the tincture of the chloride of iron, which must be administered at frequent intervals and in large doses—from 5 to 30 drops according to age and vigor of patient, should be given from every half hour to every second hour as the case may be. It is given in glycerine and water, or in diluted syrup of lemon. Dr. Cohen prescribes chlorate of potassium very frequently in this disease—in the form of *chlorine* mixture (made of an equal number of grains of the chlorate and of drops of hydrochloric acid in plain, or aromatic water, or in the infusion of quassia). He always suspends the use of this remedy when there are any symptoms of renal irritation produced by it.

He administers the hydrochlorate of quinia (in preference to the sulphate) as a tonic, antipyretic, neurotic, and antiseptic. It is to be given in decided doses. When deglutition is painful it is given by enema with proper augmentation of the dose.

Alcohol, in the form of strong wine, or as brandy or rum, is regarded as of the utmost importance when the system begins to give way. It should be given after the earliest manifestations of decided loss of vigor. At this stage it is of more importance for the time being than any remedial agent. From f ℥ ss. to f ℥ j of brandy are to be given at intervals of from fifteen minutes up to three hours. As long as it is well borne it may be given to any extent short of intoxication. Children readily take a sort of syrup of brandy made by burning it beneath a lump of sugar, which becomes melted in the process. At moments of sinking he regards carbonate of ammonium as valuable. He gives from two to ten grains by the mouth, in syrup of acacia, or from eight to forty grains by the rectum. At moments of collapse the ammonia is given by intravenous injection.

The sore throat is treated by pellets of ice, placed in the mouth and renewed more or less cautiously. The use of ice compresses is not approved. It is thought better to apply warm cotton batting, spongio-piline, or an actual cataplasum, or to anoint the neck with oil, lard or cosmoline, care being taken not to abrade the cuticle lest local infection arise as a complication. Morphia is administered when great pain arises.

Morbid products in the pharynx and nasal passages undergoing detachment should be promptly removed. This morbid product is kept diffluent as much as possible by maintaining an excess of humidity in the atmosphere of the room by keeping a steaming vessel of water on the stove. The uninvaded tissue should never be cauterized. Applications of the tincture of the chloride of iron should be made to the pseudo-membrane with a swab of cotton or sponge. After this application the attempt may be made to remove the deposit by gargle, spray-douche, or syring; employing limé water as the medium. Forcible removal of the deposit is not regarded as judicious.

When the larynx is invaded Dr. Cohen keeps a constant stream of steam in motion directed over the patient's face. Whenever the respiration becomes obstructed, a few pieces of lime about the size of the fist are slacked by the bedside every hour or so, covering the vessel in which they are slacked with a hood of stiff paper, so as to direct the steam and particles of lime towards the mouth of the patient.

The use of emetics is indicated in children to provoke expectoration from the air-passages in the act of vomiting; but the same indication does not occur in adults who are able to expectorate voluntarily. If successful, the emetic may be repeated, at intervals of six hours, as long as the indications continue to recur. Alum, ipecac and turpeth mineral are the most reliable agents, and may be tried in the order named; adhering to the alum if it prove efficient. Emesis should not be carried too far, or be repeated if ineffectual, as it exhausts the power of the system without any compensation in the discharge of morbid products.

Should asphyxia be threatened from accumulations in the larynx or trachea, tracheotomy is indicated, and, though most frequently unsuccessful in averting death, it facilitates due access of atmospheric air to the lungs, and often saves lives that would otherwise be lost. The most careful attention is required after tracheotomy to keep the artificial passage clear. The stimulating treatment and the lime inhalations should not be discontinued. The two main indications for favorable prognosis after tracheotomy are desired for food, and ability to expectorate. All treatment should be subservient to facilitating these great ends.

Dr Bartholow believes that there are two objects to be kept in view in the treatment of diphtheria:

1. To modify the course and shorten the duration of the disease; 2, to obviate the tendency to death.

First head.—The application of topical agents to the fauces and the administration of internal remedies according to symptoms.

He entirely disapproves of caustic and acid applications as inviting the disease to the adjacent portions of the mucous membrane, by destroying the epethelium. He does not think much of the value of benzoate of sodium. The application of sulphur, in the form of powder by insufflation, or by blowing it over the whole diseased surface as far as it can be reached, he believes to be as a good treatment. He regards lime water and lactic acid as of value as solvents, Some pieces of freshly burned lime are put in water and the patient directed to breathe the vapor as it rises, or a solution of lactic acid strong enough to taste distinctly sour, is freely applied to the throat by a large mop. He places no value in the use of chlorate of potassium or tincture of the chloride of iron as faucial remedies. When gangrenous sloughs are thrown off from the throat, carbolic acid is indicated, a one-per-cent. solution—not stronger than one per cent. This solution may be applied either by mop or syringe. When the exudation extends into the nares, the spray of a one-per-cent. solution of carbolic acid is gently thrown into them and kept up until the two canals are pervious, thus preventing the extension and decomposition of morbific materials and the consequent swelling of the deep cervical glands and possible development of septicæmia. It is only when the exudation extends into the nares that much good can be accomplished by topical applications—so thinks Dr. Bartholow.

Second Head.—The prevention of the diffusion of the morbific agent, of the development of septicaamia, and of failure of the heart. With the earliest appearance of an exudation in the fauces, from two to ten grains of the bromide of ammonium are given every three hours. It is believed that the diffusion of this agent through the mucous membrane of the respiratory organs and so out of the mouth, detaches the exudation. To prevent septic decomposition he advises the use of a drop or two of Lugol's solution in water every hour or two. This drug is to be given when the exudation is fully developed and spreading. He uses alcohol steadily, pushing it in large doses as an antiseptic agent. Quinia is also considered valuable in this same connection. Dr. Bartholow does not believe in the extraordinary powers of chlorate of potassium in this disease, as claimed by many. He fears its injurious effects on the kidneys.

As food, milk, egg-nog, and beef-tea are given freely about every three hours.

Dr. Abram Jacobi sums up the treatment as follows: Every case should be treated on general principles with symptomatics, roborants, stimulants, febrifuges, externally, internally, or hypodermically.

The uncertainty of the termination and the frequency of collapse, or sepsis, prohibit procrastination. Waiting long means often waiting too long. Alcohol is a very important adjuvant and remedy.

The dose must often be apparently large, from two to twelve ounces daily, according to the circumstances.

Depletion is absolutely contra-indicated. Debilitating complications, such as diarrhœa, must be stopped instantly. Stomatitis, chronic pharyngitis, hypertrophy of the tonsils, glandular enlargements must be relieved or removed preventively. Acute catarrh of the mouth and pharynx requires the use of potassium or sodium chlorate, in doses not exceeding a scruple daily for a child of a year, one to two drachms for an adult. The single doses must be small and very frequent, ever hour, half, or quarter hour. Large doses are dangerous, result often in nephritis, and have proved fatal. The main indication in local diphtheria is local disinfection. To disinfect the blood effectively we have no means. Salicylic acid changes into a salicylate which is no longer a disinfectant. The amounts of disinfectants required to destroy bacteria is so great that the living body could not endure them. But the discipline of the house, school, and social intercourse can be so modified as to prevent the spreading of an epidemic. The inhalation of steam is very useful in catarrh of the respiratory organs, and also in inflammatory and diphtheritic affections. In fibrinous tracheo-bronchitis it has proved quite successful. But it may also prove dangerous by excluding oxygen and overheating the room or tent. Drinking large quantities of water, with or without stimulants, also excites action of the muciparous gland and aids in macerating membranes. The internal use of ice, and its local application to the affected parts, can be very useful. But the cases must be selected for each and any of the remedial agents and applications. The use of baths and the cold and hot pack is controlled by general indications. The usefulness of lime-water and lactic acid has been greatly over-estimated. Glycerine is a valuable adjuvant, both internally and externally, but nothing more. Turpentine inhalations are deserving of further trials, though they are more effective in purely inflammatory than in diphtheritic processes. Inhalations of chloride of ammonium act favorably in catarrhal and inflammatory conditions, and deserve a trial for the purpose of aiding maceration of membranes. Mercurials are contra-indicated in the septic and gangrenous forms of diphtheria, but in those which assume the purely inflammatory character, with less constitutional debility and collapse, as in sporadic croup or in fibrinous tracheo-bronchitis, some reliable clinicians claim good results. Astringents, such as alum, do not work favorably. Chloride of iron is amongst the most reliable of antiseptic and astringent agents. Small doses at long intervrls are quite useless. Moderate doses frequently repeated have a satisfactory general and local effect. A child of a year must take at least a drachm daily; a child of three or four years, from two to three drachms. The same or larger doses for an adult. The chloride is to be mixed with water and glycerine, in various proportions, so that a dose is taken every hour, every half-hour, every ten minutes. Thus the local applications to the throat become almost superfluous. Potassium or sodium chlorate, half a drachm to a drachm daily, may be added with advantage. Carbolic acid is useful both in local and internal administration. According to the end to be reached, it may be used either in concentrated form or in a one per cent. solution. Internally, in doses of a few grains to half a drachm daily. Salicylic acid acts as a caustic when concentrated; in moderate solutions it destroys fetor; salicylates are anti-febrile only. The anti-febrile effects of quinia are not so favorable in infectious as in inflammatory fevers; its antiseptic action is not satisfactory in practice. Deliquescent caustics are dangerous. Injury of the healthy mucous membrane must be avoided. Mineral acids, and particularly carbolic acid, when their application can be limited to the desired locality, are preferable. Bromide, both internally and externally, is warmly recommended by Wm. H. Thompson. Boric acid, in concentrated and milder solutions, has been recommended as a local application to membranous deposits generally, and to the diphtheritic conjunctiva in particular. Membranes must not be torn off, and not removed unless they are nearly detached. Caustics are contraindicated, except where their application can be limited to the diseased surface. No healthy part must be torn. Swelled lymph glands require ice, iodine, idoform, mercury, poultices, incision, carbolic acid, according to circumstances, and at all events frequent and careful disinfection of the mucous membrane from which their irritation originates. Diphtheria of the nose is apt to be fatal, unless careful treatment is commenced at once. It consists of persistent disinfectirn of the nares and pharynx by injections. The tendency to sepsis forbids a long intermission of them. They must be continued day and night, for one to several days, no matter whether the glandular swelling be considerable or not. Laryngeal diphtheria proves fatal in almost every case, unless tracheotomy be performed. It is the less successful the more the epidemic or case bears a septic character. Emetics are useful for the removal of the half-detached membranes. Diphtheritic paralysis requires good and careful feeding—iron, strichnia, the faradic ar galvanic currents, friction, hot bathing. Urgent cases indicate the hypodermic administration of strychia. Diphtheritic conjunctivitis is benefited by ice and boracic acid ; cutaneous diphtheria by local cauterization and disinfection, besides general treatment.

CARDIAC DROPSY.

There are some cases in which there is nothing to be done but to stimulate the action of the heart. Dr. Pepper holds that in a case of œdema of the lower extremities, due solely to cardiac failure, absolute rest, small quantities of food at short intervals (because the digestion is weak,) a little alcohol and plenty of digitalis, will in a few days cause the dropsy to disappear. The great element in the treatment is absolute rest. He finds that the simple fact of making such a man sleep upon the ground-floor does more for his comfort than any drugs. It prevents him from having to lift the weight of his body in going up stairs, and also enables him to take exercise in the open air without going up or down stairs. The carrying out of this simple element in the treatment of heart failure is often followed by a disappearance of the dropsy, which seems almost miraculous. Regarding the use of food in cardiac disease with dropsy, Dr. Pepper says that in all such cases liquid food should be preferred to a solid diet, for the reason that a liquid diet stimulates the action of the skin. This is true whether strong meat broths, beef tea, or milk, are given ; for after all beef tea is nothing but a mineral water containing various salts and a little albumen in solution. In the same way milk contains a great many salts and with these albumen and sugar. Therefore a liquid diet of milk and broths is really a diuretic and diaphoretic in itself, while the elements of nutrition are also present in an acceptable form. In almost every case of cardiac disease with dropsy, a diet chiefly liquid is an important element in the treatment. This becomes more important when the digestion is weak or in any way deranged. In cases with hepatic and renal derangements it may be laid down as a rule, that a liquid diet is to be used. In case of hepatic involvement in which there is a dimished secretion of bile, it will be found that skimmed milk, or better, buttermilk, are very useful, because they contain no fat. The food should be taken at short intervals and in small quantities. A small quantity of green vegetables and lean meat will often have to be allowed, in order to satisfy the patient.

Alcohol is a very important element in the treatment. The strong forms of alcohol are injurious, but the weaker forms, as diluted sherry or gin diluted with water, which, on account of its containing juniper berry, is diuretic, are well digested and stimulate the action of the heart. Lastly, digitalis, which in this condition is the most valuable of all drugs, should be given freely, watching its effect. Twelve drops of the tincture is considered an average dose by Dr. Pepper.

In cases where, instead of there being failure of the heart, that organ is acting powerfully, and there is extreme nervous congestion, the treatment has to be changed. Here, while we use the same elements—rest and modified diet—alcohol is injurious, and digitalis does not of itself suffice. The heart, lungs, liver, and kidneys, have now to be relieved by depletion of the engorged veins. For the relief of the lungs dry cups are applied over them and saline expectorants are administered. In the case of the liver, repeated doses of mercurials, followed by a saline laxative, should be used. In no condition of the system are better results seen from repeated doses of mercurials followed by a saline laxative, than in hepatic congestion from cardiac disease. Three grains of blue mass once or twice a week, followed by a saline, will be found to be a most important and valuable element in the treatment. In the case of the kidneys, dry cupping and saline diuretics should be used. Buttermilk is also useful in this condition.

There is another question to be considered, that is, the treatment of the anæmia. Some practitioners fall into a stereotyped way of treating cardiac diseases. Some give only iron, others give only digitalis; but the only right way to treat the disease is according to the indications of each case.

Iron is useful where there is anæmia. It does not make much difference what form of iron is used, but preferably the diuretic salts, the potassi et ferri tartras, the citrates and the chlorides, should be used in cardiac dropsies. They should be given in large doses, beginning with a moderate dose and slowly increasing it.

TYPHLITIS.

Dr. Loomis treats this disease successfully by distending the colon with large enemata, the patient being placed in the knee chest position during the injection. Hot fomentations as applied over the abdomen, and opium and belladonna given to quiet the pain. This means is preferred to that of catharsis, to obviate any possibility of peritonitis, it having been found that the administration of powerful purgatives is liable to produce decided irritation of the already inflamed intestine. The enema used consists of starch and warm water. Laudanum may be added to further quiet the inflammation and pain.

THE COUGH OF PLEURISY.

In cases of dry pleurisy in Bellevue the cough is often a distressing symptom, and unlike a cough with exporation, does not relieve the patient. It is supposed that this cough is reflex, due to irritation of the plural surfaces, produced by the motion of the chest-wall. The treatment is therefore directed to the cause, and the movement of the ribs is arrested by straping the thorax with adhesive plaster on the side affected. This method of treatment often gives a surprising amount of relief, not produced by the ordinary method of counter irritation. As the cough in the initial and final stages of pleurisy with effusion, may also be due to a similar cause, the treatment is applicable in these cases also. When it is found by the patient that the cough is less when he is lying upon the affected side, thus partially immobilizing it, the indication for strapping the chest is considered positive.

THE COUGH OF PHTHISIS.

When the cough is severe and persistent, Dr. Bartholow is obliged to use opium. He thinks codeia the most useful of the derivitives and preparations of opium. He combines it with strychnia when there is vomiting, and with atropia or picrotoxine when the sweats are profuse, viz:

℞ Codiæ sulph., ... gr. x;
    Ext. hyoscyami., ... ℈j.
    M. et ft. in pil No. xx.
Sig.—One pill every four hours.

℞ Codiæ sulph., ... gr. xvj;
    Strychniæ sulph., ... gr. j;
    Atropiæ sulph., ... gr. ⅛;
    Acid. sulph. dil., ... f ʒij;
    Aquæ, ... f ʒvi. M.
Sig.—Ten to fifteen drops thrice daily.

Morphia may be substituted for codeia in either of the above by reducing the quantity one-half.

Carbolic acid he has found very serviceable. He uses this formula:

℞ Acid. carbol., ... gr. viij;
    Aquæ lauro-cerasi.
    Aquæ, ... āā f ʒj. M.
Sig.—A teaspoonful every four hours.

Carbolic acid is especially indicated in the fetid expectoration of bronchiectasis.

In the cough of fibroid lung he uses this formula:

℞ Ammonii iodidi, ... ʒij.
    Vini picis liquidi, ...
    Syrup tolu., ... āā f ℥j. M.
Sig.—A teaspoonful at a dose.

CROUPOUS DYSENTERY.

Opium is given in sufficient quantity to keep the patient comfortable. He is fed with liquid food, and as soon as the heart's action begins to grow feeble stimulants are freely administered. Dr. Delafield thinks it is important in these cases not to pay too much attention to the local symptom. From time to time it is necessary to give a laxative in order to empty the colon of its fæcal contents, for the patient's small, bloody, and mucous passages are not, as a rule, fæcal passages.

The two drugs which are used for the purpose of acting directly upon the inflamed condition of the bowels, are ipecac and calomel. The dose of ipecac is gr. x-xxx. Opium in small quantities should be combined with it, and the patient kept quiet in bed to prevent vomiting. This dose of ipecac may be repeated two or three times in the twenty-four hours, until as many as six doses have been taken. The dose of calomel given for the same purpose is gr. x-xx. This large dose of calomel should not be repeated more than twice for fear of salivating the patient This calomel treatment has proved less successful than that by ipecac in Dr. Delafield's experience, but still there are occasional cases in which it does seem to be of service.

RHEUMATISM.

Dr. T. G. Thomas calls attention to a combination of salicylic acid which he has used many times with good results in both acute and subacute rheumatism, as well as in a few chronic cases of the disease. For this combination he claims the following advantages: that it does not disturb the digestive system; that it is very palatable; that it forms a perfect solution of salicylic acid; that it is effective in curing the disease; that it produces no bad effects upon the heart; and that it is less depressing than salicylate of soda. The formula is as follows:

℞ Potass. acetat., . . , . . . ℥ ii.
   Acid. salicyl., ... ℥ ss.
   Aq. menth. pip., ... f ℥ iv.
   Syrup. limon., ... f ℥ ij. M

It is best prepared by placing the potash and peppermint water in a porcelain mortar and gradually adding the acid, triturating to a perfect solution, and then stirring in the syrup. The dose is a teaspoonful every two, three, or four hours, or oftener, according to the violence of the attack. This dose gives twenty grains of the acid to eighty grains of the acetate. In the robust class of patients without complications, Dr. Thomas relies exclusively upon it, with an occasional hypodermic dose of one-sixtieth to one-eightieth of a grain of atropia, or combined with morphia in cases where the atropia alone is insufficient to allay the pain; such patients are usually convalescent in five or six days.

THE DIARRHŒA OF PHTHISIS.

Rest in bed, with absolute liquid diet of the most nutritious character, consisting of milk, arrowroot, meat broths, finely-minced meat, and other articles of this kind, must be continued until the tendency has been entirely overcome and the stools have been normal for a day or two. A moderate amount of opium must be used to control the tendency to diarrhœa. As to astringents, nitrate of silver, given in minute doses, guarded by opium, is, according to

Dr. Pepper, a most valuable remedy. Bismuth does well in most cases. Sugar of lead or tannic acid with opium is also very good. Aromatic sulphuric acid with sulphate of morphia and camphor solution will do very well in many cases. The nitrate of silver may be given in pill form, or in small doses dissolved in the syrup of acacia. If there is a wide-spread catarrh, with involvement of the duodenum, and extension of irritation into the bile ducts with obstruction to the flow of the bile, the treatment must be prefaced, by small doses of calomel, soda and opium, given for a few days until the tongue begins to clean and the character of the stools shows that the bile passes freely. Then nitrate of silver and opium may be substituted, or one of the other astringents. Such patients are very susceptible to the action of mercury, and the calomel must be used in small doses. It may be given according to the following formula:

℞ Hydrarg. chloridi mitis, ... gr. ij.
   Sodii bicarb., ... ℨj.
   Pulv. opii., ... gr. iij.
   Pulv. No. xij.   M. et ft.
Sig.—One every four hours.

These are to be taken until they produce the desired effect, or until all are taken.

ACUTE BRONCHITIS IN CHILDREN.

℞ Tc. verat. vir., ... ♏xii;
   Syrup. scillæ comp., ... f ℨii;
   Syrup. balsam. tolu., ... f ℨiv. M.
Sig.—One teaspoonful every two or three hours to a child five years old, in the first stages of the disease.

When the temperature falls and moisture appears on the skin, under the influence of the above prescription, Dr. J. Lewis Smith stops its administration and resorts to such expectorant mixtures as this:

℞ Ex. cubeb. fl., ... ♏xl-f ℨi;
   Syr. simp. . . . . f ℥ij ss.
Sig.—A teaspoonful three or four times daily.

THE LARYNGITIS OF PHTHISIS.

In the early stages, counter-irritation, by means of a blister, or by the repeated application of iodine, or of iodine and croton oil, and the direct application of astringents, as sulphate of zinc, grs. x-xl to the ounce, will often cut short the attack. If the attack is more marked, the larynx must be put to rest as much as possible, by inducing the patient to do without speaking, not even whispering, and to communicate everything by writing. The food should be unirritating, and great care should be taken that no particles get into the larynx. The patient should be careful that he does not take a fresh attack from chilling of the surface. Opium must be given, to allay irritability and cough. Counter-irritation must be steadily kept up. Topical applications must be used, to relieve the swelling and favor cicatrization of the ulcers.

Now as to topical applications. Dr. Pepper thinks that in treating these cases the strength of the solutions used is often too great. Mild applications will do more good than strong ones. Solution of sulphate of zinc, or of the salts of iron, or of iodine, will be found of service. Lately he has been using a solution of iodoform in ether or chloroform, varying in strength from gr. xv to the ounce up to a saturated solution, which is about ʒj to the ounce. This is a very unirritating application, and it certainly possesses very marked absorbifacient and alterative powers. Of course, each of these applications is to be made only to the diseased spots by the aid of the laryngoscope.

Inhalations are very valuable in the treatment of the laryngeal complication of phthisis. They may be used with the ordinary atomizer, which is a good method, but apt to excite coughing. In this way we may use nitrate of silver, tannic acid, sulphate of copper, sulphate of zinc, iodine, etc. We may use an inhaler made of a piece of rubber tubing, in which is placed a number of small rolls of bibulous paper. The solution to be inhaled is dropped in at one end, and then the patient uses it as an ordinary cigar. He has made a little alteration in this, by substituting for the rubber tube a glass tube with a constriction near one end, and for the paper, pumice stone, which makes it more cleanly. With this we may use a number of volatile solutions. In other cases we may employ a sort of mask made of two pieces of rubber, with sponge between them. By this means the patient may breathe medicated vapors for hours at a time. We may use with this tar, carbolic acid, iodine, etc.

NOCTURNAL INCONTINENCE OF CHILDREN.

Dr. S. D. Gross has discovered that this prescription will speedily relieve the irritability of the bladder, especially if conjoined with such means as a cold shower bath daily, the avoidance of irritant food and late suppers. The patient must sleep on the side or belly, and take care to drink nothing for the few hours preceding sleep, and to empty the bladder on going to bed.

℞   Strychniæ, ... gr. ⅓ ;
    Pulv. canthar., ... gr. ij ;
    Morph. sulph., ... gr. jss ;
    Pulv. ferri., ... ℈ j.
    M. et ft. pil. No x x.
Sig.—One thrice daily to a child ten years old.

THE HÆMOPTYSIS OF PHTHISIS.

An essential element of Dr. Pepper's treatment is absolute and perfect rest in bed, avoidance of talking, and the suppression of cough, as far as possible. Then, everything which renders respiration easier should be encouraged. Nothing should obstruct the neck and chest. The room should be cool, the head and shoulders elevated, and the patient should swallow little pieces of ice. He thinks that the cold exercises some influence in checking the hemorrhage. All the nourishment should be cold and non-stimulating, as, for instance, cold milk and meat broths. As to the local application of cold to the outside of the chest, he does not think that this is is advisable. It is probably true that if we could localize the seat of the hemorrhage, and if it was superficial, the cold pack might aid in checking it, but he thinks that the risk of the bad effects from the cold would be too great. He prefers to resort to dry cupping over the back and front of the chest, over the spot from which the blood comes, if this can be located ; but where you cannot locate the exact seat, a dozen dry cups may be applied over the back and front of the chest, if it can be done without disturbing the patient too much.

Of drugs, ergot seems to be the most powerful in checking hæmoptysis. The extractum ergotæ fluidum may be given in doses of a teaspoonful every fifteen minutes, until the hemorrhage is stopped, and then continued in smaller doses, or it may be given by hypodermic injection in doses of gtt. xv, or ergotine may be used. If the stomach is irritable, gr. v of ergotine may be given per rectum. Sometimes ergot will have no appreciable effect. Under such circumstances gallic acid is the next best remedy. He frequently combines it with aromatic sulphuric acid, which makes a more efficient and pleasant mixture.

℞ Acidi gallici, ... ℨij ;
Acidi sulphurici aromat., ... f ℨj ;
Glycerinæ, ... f ℨj ;
Aquæ, q. s., ut. ft., ... f ℥ vj. M.
Sig.—A tablepoonful, as required.

This is to be given every hour, every half hour, or at shorter intervals, until the hemorrhage is brought under control. This ranks next to ergot, and where the stomach refuses ergot, or where ergot produces no effect, he usually resorts to this combination.

Opium should be given to control the cough, quie nervous irritability, and allay vascular excitability. The opium may be given either by suppository, by hypodermic injection of morphia, or the deodorized tincture may be added to the ergot or gallic acid combination. For internal use there is no preparation of opium equal to the deodorized tincture, the tinctura opii deodorata. Where the stomach is irritable, suppositories of opium or hypodermic injections of morphia may be used. The amount must be determined by the effects, but persons with hemorrhage will bear large quantities of opium. The proper method is to give small doses, frequently repeated, until the desired effect is produced.

THE GASTRIC DISTURBANCES OF INITIAL PHTHISIS.

Dr. Bartholow puts the patient on milk for a week or two, and then gradually constructs a suitable dietary. To remove the condition of gastro-intestinal catarrh there are several remedies on which he relies with confidence. Arsenic stands first, and is to be given in small doses—two drops three times a day, before meals. Scarcely inferior to arsenic is the combination of iodine and carbolic acid.

℞ Tinct. iodinii, ... ℥ j.
    Acid. carbol.. . . . . . . ℥ ss. M.
Sig.—One to two drops in a tablespoonful of water three times a day, before meals.

The oxide and nitrate of silver are also highly serviceable but, unfortunately, they cannot be continued long, owing to the danger of argyria. The mineral acids, if the digestion merely flags, or if the previous remedies have been taken for a sufficient time, are very important remedies. He has seen some striking results from the use of the nitromuriatic acid, during the initial stage, but the action of this remedy in changing morbid states of the mucous membrane, and in promoting appetite and digestion, is the secret of any curative power it possesses. The alkaloid strychnia, dissolved in diluted muriatic acid, makes a useful combination, for strychnia is probably the most effective agent we possess for arresting the vomiting of phthisis.

If the morbid state of the mucous membrane has been removed, and the appetite restored, he then considers the question of forcing the nutrition by increasing the performance of the assimilative organs. He now gives cod liver oil, and aids its digestion by the administration of malt liquors or alcohol in same form. It should be given after meals, to be digested and assimilated with the other foods. It is a great error to administer the oil, as is so often done, on an empty stomach and sometimes before the meal, because it will spoil the appetite, and keep the stomach at work during its proper period of repose.

Furthermore, the quantity of oil taken should rarely exceed a teaspoonful, for this amount only can the digestive organs properly assimulate at one time. If it impair the appetite and cannot be digested it is useless, and should not be continued. Bernard has put us in possession of an important fact, which ought always to be utilized—that is, that ether promotes digestion of the oil when administered with it. Bernard's notion was that ether increases the flow of the pancreatic juice, and must therefore promote the emulsionizing of the oil. Whether or no the theory be correct, there has been accumulated sufficient testimony to show that the digestion of the oil is much more easily and perfectly accomplished by the addition of some minims of ether.

To what extent is the administration of alcohol desirable and proper in the treatment of phthisis? First of all, we should take our position firmly against a prevalent notion that whiskey is in a certain degree antidotal, and that any case of consumption must be cured if only it is given early enough, and in sufficient quantity. This mischievous fallacy is so far from true, that we now know a form of degeneration of the pulmonary parenchyma is produced by alcoholic excess. Before hectic comes on, as a rule, the extract of malt and malt liquors are preferable to whiskey and brandy. When wasting proceeds rapidly, and destruction of the lung tissue is going on at the same pace, when the fever approaches more and more the scepticæmic type, the stronger liquors are better, but at any period whiskey should be given instead of malt liquors, if it agrees better. Is there any guide to the proper administration of alcohol? He is perfectly clear that these are well-defined principles. That quantity of spirit which increases the appetite and improves digestion is the proper quantity. An excess is hurtful, because the alcohol precipitates the pepsin, from its solution in the gastric juice, and therefore suspends the production of peptones. A large quantity taken on an empty stomach is without influence on the digestion of foods, and whilst it affects the mucous membrane injuriously, after absorption, acts on the hepatic cells. These reasons seem conclusive in favor of a moderate quantity of alcoholic food—say half to an ounce of whiskey—taken after each meal.

NERVOUS COUGH.

Dr. Bartholow has found this formula most successful in treating the cases of cough by habit after the cessation of the whooping cough proper. It is also very useful in allaying the nervous cough of mothers, which exists during the presence of cough in the household:

℞ Acid. hydrocyan. dil., ... gtt. xvi;
   Tc. sanguinariæ, ... f ʒiv;
   Syrup senegæ, ...
   Syrup tolu., ... āā f ʒvi;
   Aquæ lauro-cerasi, q. s. ad., . . . f ʒiii. M.
Sig—One or two teaspoonsfuls, according to age, every 3 or 4 hours.

THE NIGHT SWEATS OF PHTHISIS.

The removal of the patient to a cool and bracing climate greatly lessens these sweats. A diet consisting largely of liquid food (the liquid acting as a diuretic) often does good. Cool bathing or cool sponge bathing with water containing salt and alcohol, are often followed by admirable results in

Dr. Pepper's hands. The surface of the body is rapidly sponged and quckly dried by friction with a towel. Among medicines atropia given at bedtime in doses of gr. $\frac{1}{100} - \frac{1}{60}$ is most powerful. Locally quinia with gallic acid or acetate of lead may be employed. Sometimes night sweats can be checked by the external use of hot water containing alum or alcohol.

IRITIS.

This formula is Dr. Bartholow's:

℞ Morph. sulph., ... gr. iv;
    Zinci sulph., ... gr. iij;
    Atropiæ sulph., ... gr. ii;
    Aquæ distillat., ... f ℨj. M.
Sig.—To be employed as a lotion.

THE FEVER OF PHTHISIS.

If any malarial element be present, Dr. Pepper removes it by the continued use of quinia and arsenic. He regards the latter as a very valuable remedy, but it must be given with care. If it is well borne its use must be kept up for a long time. The proper dose of quinia is from gr. xii-xv.

Where the grade of hectic is lower and is due mainly to the local irritation in the lungs, the dose of quinia should not be more than gr. i—ii daily. When the pulse is rapid and weak, digitalis is associated with the quinia. On the other hand where the system is relaxed and a tendency to night sweats present and the pulse not rapid, aromatic sulphuric acid is given with the quinia. In nearly all these cases of hectic fever a little opium may be used with advantage.

This is a good formula for a high degree of nervous irritation and hectic fever:

℞ Quiniæ, ... gr. x;
    Pulv. dig., ... gr. iv;
    " opii, ... gr. ij;
    M. et. in pil. No. x div.
Sig.—One thrice daily. Sometimes arsenious acid may be added with benefit.

This for marked tendency to relaxation of system and night sweats:

℞ Quiniæ, ... gr. x;
   Acid sulph. aromat., ... f ʒij;
   Pulv. opii, ... gr. ij;
   Aquæ, q. s. ad., ... f ℥ v. M.
Sig.—A tablespoonful thrice daily.

CONSTIPATION.

Where constipation is due to torpor of the muscular layer of the intestine, combined with defective secretion of the mucous membrane, Dr. Bartholow uses either one of these formulæ:

℞ Tc. nucis vomicæ;
   Tc. belladonæ;
   Tc. physostigmæ, ... āā f ʒij. M.
Sig.—Thirty drops in water, morning and evening.

Or ℞ Ex. physostigmæ;
   Ex. belladonnæ;
   Ex. nucis vomicæ, ... āā gr. v.
   M. et. ft. in pil. No. x.
Sig.—One pill at bed time.

ROUND WORM AND ASCARIS VERMICULARIS.

For the expulsion of the round worm no better formulæ than these have been designed. Dr. J. Lewis Smith has also found them an effectual means of destroying the ascaris vermicularis.

℞ Fl. ex. spigeliæ, ... f ℥i;
   Fl. ex. sennæ, ... f ℥ss; M.
Sig.—A teaspoonful for a child 3-5 years old.

℞ Fl. ex. spigeliæ;
   Fl. ex. sennæ, ... āā f ʒ iv;
   Santonini, ... gr. viij. M.
Sig.—A teaspoonful to a child of five.

CARDIAC HYPERTROPHY.

Dr. Alonzo Clark does not sanction bleeding in this connection. All that can be done in his opinion is to induce the patient to take as much food as he can digest, and to administer some form of iron. All kinds of violent exercise are strictly prohibited.

FATTY HEART.

The patient is directed to take moderate exercise. All fatty food, including butter, milk, cream, and fat of meats, is prohibited. The food eaten is to consist of lean beef, mutton, chicken, and the vegetables which contain the least oily matter. As medicine the bicarbonate of soda is administered by Dr. Alonzo Clark, but never to the point of rendering the urine alkaline, lest urinary calculi be formed.

SURGICAL AND VENEREAL DISEASES.

TREATMENT OF HÆMORRHAGE.

Dr. A. L. Ranney gives the following concise rules for meeting all possible indications in the treatment of hæmorrhage:

(1.) Always ligate the bleeding vessel in moderate hæmorrhage when convenient to do so. (2.) Use compression over the wound on the main trunk in moderate hæmorrhage when ligature of the wounded artery is inconvenient. (3) In violent hæmorrhage enlarge the wound and tie the artery. (4.) as a rule, never attempt ligation except when bleeding actually exists. The exceptions to this rule are: (*a*) in exposed vessels of large calibre demanding ligature as a safety measure; (*b*) in delirium tremens following an injury; (*c*) when necessity for transportation exists. (5.) Ligation should, as a rule, be applied at the bleeding point, and not remote from it. (6.) Use the external wound as a guide to your incision to reach the vessel, except when the wound exists on the side opposite to the vessel injured, when a probe may be cut down upon. (7.) Always use the greatest precaution to avoid needless loss of blood in reaching the vessel until the fingers can compress it. (8.) The artery when found should be tied above and below the wounded portion, and at a bifurcation three ligatures should be used. In case the lower end cannot be discovered, use compression in the wound as a substitute for ligature. (9.) A ligature should not be placed close below a large branch. (10.) In recurring hæmorrhages the treatment should depend on the color of the blood and on the severity of hæmorrhages. If the hæmorrhage springs from the proximal end of the artery: (*a*) tie if possible;

(*b*) amputate if necessary; (*c*) use styptics and compression if both are possible. (11.) Amputation is preferable to ligature: (*a*) when great swelling of the limb renders ligation difficult; (*b*) when exhaustion of the patient forbids further search for the vessel; (*c*) when competent assistance is needed and not attainable. (12.) In case a large vessel is injured without actual hæmorrhage, hot flannels to the limb are indicated as a preventive measure. (13.) In case an aneurism is the seat of the hæmorrhage,—provided the aneurism is traumatic in its origin,—it should be treated on the same principles as if it were a wounded artery.

HYGIENE OF THE MOUTH IN SYPHILIS.

Dr. E. L. Keyes says that on account of the necessity of giving mercury, and of the danger of salivation, lesions of the mouth and throat, which are very obstinate in this disease, should be avoided so far as possible, by cleanliness of the mouth and freedom from irritants. Before the mercurial course is commenced let the teeth be put in order by a dentist; let any sharp angles of the teeth likely to come in contact with the tongue be filed away. Any reaccumulation of tartar during the progress of treatment should be removed. Let a soft tooth-brush be used. The tooth-powder should be strongly alkaline and slightly astringent. A half-teaspoonful of bicarbonate of soda and a teaspoonful of the tincture of myrrh, in a glass of water, or white castile soap and water, or a weak solution of alum in water make excellent toothwashes. With such care mucous patches become less annoying and easier to manage, and the effect of the mercury may be more closely watched, since one is not apt to be misinformed as to the cause if the edges of the gums become soft and tender. Smoking is also entirely contra-indicated during the first year or two of syphilis, as it is apt to induce a greater number of mucous patches and mouth lesions than would otherwise occur. Tobacco-chewing is equally bad, or worse. Highly spiced or stimulating food may help to keep the mouth tender, and should therefore be avoided. A pipe is a dangerous thing in syphilis, owing to the risk of infection, if it is used by healthy persons, because the secretions of mucous patches and syphilic ulcers in the mouth are specially contagious.

SPINA BIFIDA.

Dr. Lewis A. Sayre says that the object of mechanical treatment is simply to protect the parts from all pressure and all possible injury until the process of ossification is completed throughout the entire length of the spinal column. This he accomplishes by first slipping over the trunk a tightly-fitting knit-shirt, similar to that used in applying the plaster jacket in Pott's disease or lateral curvature. Then, having the patient held in a firm position, but without being suspended, he passes a few turns of a plaster bandage around the trunk and pelvis in such a manner as to cover the spina bifida completely. After this he cuts off a piece from both the top and bottom of the shirt. He then makes a few more turns of the plaster bandage outside of all, and finally, before the plaster has had time to set, presses in the plaster with his hands on both sides of the tumor, so as to make the covering more cup-shaped, and thus protect it the more completely from all pressure. He them makes a hard, artificial roof for the spinal cord and nerves, which takes the place of the normal bony one until nature supplies the deficiency. If, on account of the child's growth, other similar plaster casings are required, they can be applied in the same manner. He puts the child on a course of phosphate of lime, with a view of increasing the earthy phosphates in its system, and thus facilitating the further ossification of the spinal column.

CYSTORRHAGIA.

To prevent such hæmorrhage, Dr. Gouley recommends that an over-distended bladder be emptied very gradually.

Draw off 8 or 10 ounces through a soft rubber catheter during five minutes, then wait an hour before drawing any more, and so continue even if 12 to 24 hours are consumed in emptying the bladder entirely. If the urine is very offensive, draw off 10 ounces and inject an equal amount of warm water containing a scruple of bicarb. of soda. Then draw 10 ounces of the contents of the bladder, and again inject the solution and so continue until the contents of the bladder are no longer fœtid. Then proceed to empty it gradually, but never allow the over-distended bladder to be evacuated at a single catheterism. If cystorrhagia occurs, Dr. Gouley recommends the following treatment: After all the urine has been drawn, introduce a soft catheter and leave it in the bladder for 24 hours, thus allowing perfect drainage. The drainage allows the muscular wall to contract firmly, and the contraction alone will stop the hemorrhage in most cases. If the hemorrhage is severe and continuous, it will be necessary to draw off the clots by suction made with an aspirator attached to the catheter. Before doing this inject a warm solution of borax and only withdraw the same amount that has been injected. As soon as the clots are removed the bladder will contract and the hemorrhage will cease. Injections of tannin or alum are only to be used as a last resort, as they may set up a general cystitis. The administration of the fluid ext. of ergot by the mouth will aid in the arrest of severe hemorrhage, but it is needless in mild cases where simple drainage will suffice.

INFANTILE SYPHILIS.

Dr. R. W. Taylor uses:—

R Hydrarg. bichlor. . . . . gr. i;
    Potass. iod. . . . . ʒi;
    Syrup sarsanii. . . . .
    Aquæ, ... aā f ʒii. M.

Sig.—5 drops for a child two months old. Increase to 15 or 20 drops if the disease does not yield.

This formula is highly efficient in effecting a cure.

SLIGHT WOUNDS.

In dressing slight wounds, simple cerate or vaseline have been in common use at Bellevue Hospital for some time. Of late, however, carbolized oil has been substituted for these, and has given great satisfaction. The proportion of carbolic acid to olive oil is one to sixteen, and it is found that the two will mix readily, and the effect of the acid is permanent, since it does not evaporate from the oil as it will from water. Its local action is soothing, both from the bland nature of the oil and the slightly anæsthetic effect of the acid. It makes a clean dressing when applied to the surface of a wound upon a piece of lint or oakum, and not only keeps the wound clean, but at the same time disinfects it.

ULCERATING SURFACES AND ABSCESSES.

In ulcerating surfaces and abscesses which show no tendency to close, the treatment at Bellevue is directed toward the production of healthy granulations. This is often difficult to accomplish, and various applications are employed to stimulate the surface. Balsam of Peru is used largely in the hospital for this purpose. Where this does not suc- ceed the following combination is usually resorted to:

℞ Balsam .... f℥ iss.
  Glycerine .... f℥ iss.
  Camphor .... ℈ j.
  Balsam Peru ..... ℥ v.
S.—Stimulating lotion.

Oxide of zinc may be substituted for the camphor in the mixture. This is applied to the surface, or injected into the abscess after the latter has been thoroughly washed out with carbolized water once a day, and rarely fails to stimulate the growth of healthy granulation tissue. Where abscess sets in a wound, a wound, and there is danger of cellulitis, the parts are kept moist by frequent application of lint wet with lead and opium wash, or this is more effeccious than any other lotion. The proportions of the lotion are one ounce of tincture of opium to half an ounce of subacetate of lead in a pint of water.

POTTS' FRACTURE.

The plaster of Paris dressing is that which is adopted almost exclusively by Dr. Erskine Mason, both in hospital and private practice, in the treatment of fractures of the leg, whether they be simple or whether they be compound. The plaster has taken the place of the side splints which were, or are, made out of any metal you please; wire-gauze, tin, or, it may be paste board, leather, or gutta-percha. They all make a capital splint, but this is also handy, and so much better than the others that it has taken their place, although it requires more care in its application than do other splints. There is more liability of damage to the limb when using the plaster dressing than with any other. You may apply your bandage too tight, and may then go away leaving your patient, and come back to find gangrene of the extremity setting in from loss of circulation. No one, therefore, should use the plaster of Paris in his private practice until he has had considerable experience in its manipulation. If the dressing causes intense pain and swelling of the parts not covered by it, the patient should always loosen it. This can be done very easily by cutting up the plaster splint with a knife along the crest of the tibia from one end to the other. It must then be sprung apart so as to take off all pressure from the limb. If this is done it is not necessary to remove the plaster splint. It can be bound together the next day by simply applying an ordinary roller around the bandage. A very good way of telling at first whether you have drawn the bandage too tight, is to observe whether the circulation in the toes is interfered with. If upon pressing the nail a few moments, and then removing the pressure, the circulation does not return, but the parts remain perfectly white, it may be taken for granted that the circulation is being interfered with, and that the dressing has consequently been applied too tightly.

As the plaster of Paris hardens very rapidly, the patient may be allowed to get out of bed and sit up, with the foot on a chair, the day following its application. If he be a strong person he may even be allowed to go about on crutches, provided, of course, the limb does not touch the floor and bear the weight of the body.

SCROFULA.

Dr. Fessenden N. Otis finds this an exceedingly valuable preparation:

℞ Iodinii, ... gr. xxiv;
   Potas. iod., ... ʒj;
   Aquæ distil., ... f ℥ii;
   Dissolve, and add of Stewart's syrup, . f ℥ij.

Sig.—From a dessert to a tablespoonful three or four times a day.

SYPHILIS.

Dr. Keyes is most partial to the proto-iodide of mercury in one-sixth grain doses. Commencing with one granule three times a day, he gradually increases the dose until an attack of diarrhœa is produced, with pain in the intestines, or a mercurial fetor in the breath, or until a livid hue becomes faintly visible along the edge of the gums, or until the teeth themselves become a little sensitive on being snapped sharply together, or until the saliva flows more freely. When either of the above symptoms occur, the patient has reached the full dose, a dose which he may continue to take with the aid of selected food and of a little opium in most cases, without becoming salivated. This is not the tonic dose. If continued, the patient suffers in the quality of blood, and in the diminution of his physical powers. This is the specific dose, and fully exerts the antagonistic influence to syphilis. This full dose, the size of which varies in different individuals, may be continued during the active manifestations of the disease. When these decline, it should be substituted by one-half the amount, which is the tonic dose, and which may be continued steadily without injury during several years. Dr. Keyes, however, uses one-sixth instead of one-half the full dose, and continues this for from two and one-half to three and one-half years, or during six months, or, better still, a year after the last manifestation of the disease.

BUBO.

Dr. Keyes advocates the following: From the moment a slight stiffness in the groin is felt, perfect rest in bed should be insisted upon, or at least the patient should be kept off his feet as much as possible. The diet should be mild, and stimulants should be avoided. A laxative is often indicated. The chancroid should be cauterized at once if suitable for that treatment. In this way the bubo may be saved from becoming virulent. If the patient be full-blooded, bitter-water may be given every morning. Avoid leeches and all counter-irritants and vesicants, as these are liable to create a surface for specific ulceration. In case the bubo prove virulent in the earlier stage, iodine and all ointments do more harm than good. An application to the skin over the inflamed gland, of equal parts of tincture of aconite root and tincture of belladonna, made several times a day, is frequently of use. This application, if irritating, should be diluted with water. Cold applications are useful. Avoid heat unless suppuration be anticipated. If the bubo has slowly advanced, it is not virulent and the pus may, in some instances, be absorbed. Employ rest and slight pressure constantly applied to the gland. The local use of the compound tincture of iodine and water, equal parts, is under these circumstances, of value. If the abscess makes the skin tense, and its history proves it to be non-virulent, it may be evacuated by a fine aspiration needle thrust obliquely into the cavity of the abscess. By these means, with good food, cod liver oil and tonics, a bubo which has suppurated, may sometimes be absorbed, and the scar may be avoided. When the abscess forms rapidly, a poultice may be applied, suppuration hastened, and a free opening be promptly made. If the bubo has been simple, the cavity fills and heals rapidly; if virulent, the cavity becomes chancroidal and requires treatment as such.

OLD SPRAINS.

Cases of old sprains of the ankle or wrist occasionally give physicians some trouble, as the pain and stiffness remain for a considerable time. In such cases Dr. Smith considers that rest is the chief object to be secured by treatment. Liniments are often recommended, but are of little use. Absolute immobility of the joint for some time must be obtained. For this purpose he prefers the glass dressing made with the salicate of soda. The application is a simple one. The joint is bandaged lightly with a soft flannel bandage, over which a single layer of cotton bandage is applied. A thick coating of the silicate is then put on with a brush, it having been warmed, and by this means made fluid before its application. Another cotton bandage is placed over the first layer of the silicate and a second layer applied over this. The time required for the firm setting of the dressing is several hours, as this is effected by the action of the carbonic acid in the air during the process of cooling. Hence it is only applicable to those cases where the limb remains in its natural position, and cannot be used where extension is to be maintained. Its advantages over plaster of Paris and starch bandages are its absolute hardness, which is equal to that of glass; its great strength, and its light weight. These qualities, and especially the last, have commanded it in such cases as required long continued immobilization. It is not to be used immediately after a sprain, as the swelling must be allowed to subside before it can be firmly applied. It is in chronic cases that it gives greatest satisfaction, by preventing any motion of the overstretched ligaments, and by rest securing recovery.

THE CHANCROIDAL ULCER.

Dr. Sturgis arrests the virulent and destructive character of the ulcer either by the actual cautery, or by caustics or alterative applications in light cases. Of the first division of remedies the *white iron* or the *galvano-cautery* takes the front rank as a destructive agent; next to that comes the strong *sulphuric acid;* third, chemically pure *nitric acid*, and, fourth, pure *carbolic acid.* A neat way of using the sulphuric acid is the way known as Ricord's *carbo-sulphuric paste*, which is made by taking a small quantity of finely powdered *willow charcoal*, and adding, drop by drop, enough of the acid to make a paste of the consistence of thick cream. This is put on with a porcelain or glass spatula, *taking care* to carry the agent into sound tissue both underneath and on the surface of the edges of the chancroid. Nitric, or carbolic acid may be used in the same way. The advantage of this method is, that besides destroying the virulent ulcer, it makes a firm dressing by the drying of the charcoal on evaporation of the acid, which, on dropping off at the end of several days, reveals the chancroid almost, if not entirely healed up. If the acid is to be used in the fluid form, then some subsequent dressing must be employed. Of all dressing Dr. Sturgis infinitely prefers the dry to the wet. One of the best preparations is iodoform *finely* powdered either alone, or in combination, thus:

R Pulv. iodoformi, ... 1 part;
    Lycopodii, ... 2 parts.
Triturate well and apply locally.

The lycopodium has probably only a mechanical action, but it absorbs fluid very readily, while the iodoform acts as a local stimulant and alterative. Another of Dr. Sturgis' favorite prescriptions is:

℞ Pulv. iodoformi;
Pulv. acid tannici p. œ.
Triturate and use locally.

This is more astringent than the other.
Number 3 is more useful when the ulcer looks flabby or indolent, it is:

℞ Pulv. iodoformi, ... ʒi ;
Zinci sulphatis, ... gr. v ;
Pulv. acid. tannici, ... ʒi. M.
Triturate for local use.

One serious objection to iodoform in private practice is the strong and pungent smell which it has. Many attempts have been made to overcome this, and Dr. Bronson, of New York, speaks highly of combining the iodoform with some essential oil, such as peppermint, rosemary and the like, which he claims overcomes the odor without interfering with the alterative action of the drug.

When a wet, in preference to a dry, dressing is to be used, this formula of Dr. Sturgis' is an excellent one:

℞ Acid carbolici cryst., ... ʒi ;
Aquæ, ... f ℥viii. M.

Or ℞ Zinci sulph., ... gr. v-xx ;
Aquæ, ... f ℥ij. M.

This latter is an excellent dressing where the ulcer looks flabby and indolent. The strength of 20 gr. to f ℥ij should be used only when the ulcer is attended with inflammation.

Another very excellent dressing for chancroids is a weak solution of nitric acid, thus:

℞ Acid nitrici, ... f ʒfs ;
Aquæ, ... f ℥viii. M.

CARIES OF THE ANKLE JOINT.

This is Dr. T. E. Satterthwaite's treatment: If the joint is inflamed, entire rest is ordered ; if an abscess forms, it is opened ; if loose bone be detected, it is simply removed as if it were a foreign body interfering with the process of healing; if, in the subsequent process of the case, mal-position of the parts is found, a support, or brace is given to rectify the deformity.

CYSTITIS AND ENLARGED PROSTATE.

Dr. Ashhurst has recently treated a case of cystitis with enlarged prostate and recto-vesical fistula by Sir Henry Thompson's method of introducing a tube into the bladder through an opening above the pubes. The patient had for several months suffered from inflammation of the bladder, associated with great enlargement of the prostate. There was no retention of urine, but, on the contrary, the bladder was much contracted, the patient micturating frequently and with intense pain, and the catheter not bringing away more than a fluid ounce of very offensive and dark-colored urine. The patient declared that there was an opening between the bowels and the bladder, but this was doubted by the doctor. Internal administration of ergot and chlorate of potassium having failed to give relief, Sir Henry Thompson's operation of establishing a direct communication with the bladder above the pubes was performed. The only point in which the operation differed was in the use of a metallic instead of a flexible tube. When the patient was visited in the ward, about an hour after the operation, it was found that fecal matter was mixed with the urine which flowed through the vesical tube, thus showing that the patient's suspicion of an abnormal communication between the bladder and gut had been well founded. The operation was followed by no constitutional disturbance, and there has been measurable relief from pain The enormous size of the prostate prevents any attempt at relieving the recto-vesical fistula by operative means.

HEMORRHOIDS.

External Piles.—In debilitated subjects the constitutional treatment adopted by Dr. J. Williston Wright consists in the use of iron, Peruvian bark and cod liver oil. As a local stimulant to the sphincter muscles, equal parts of the confection of black pepper, the confection of senna, and the confection of sulphur, are rubbed down into a semi-liquid mass with a little honey or treacle, and administered in one or two teaspoonful doses every morning before eating. If the constipation is obstinate the dose is repeated at bedtime.

In the case of a plethoric patient, Dr. Wright insists in the first place upon a reduction of diet, which must consist mainly of vegetables and fruit. Such patients are recommended to take a large full wine glass of Hunyadi-Janos water every morning. Where mineral waters are not well borne, equal parts of the carbonate and sulphate of magnesia, precipitated sulphur and the bitartrate of potassium, are rubbed together in the form of a powder and given in doses of one, two, or three teaspoonsful in sweetened water, or in syrup, in the morning. The best local treatment of external piles consists in removal, but where the patient will not consent to an operation, the parts are very thoroughly smeared with a mixture of equal parts of the extract of belladonna and the aqueous extract of opium, rubbed down into the form of a fluid, with a little glycerine. This is followed by the application of a warm poultice well saturated with the tincture of opium. The patient is then put to bed, with his head low and two pillows are put under his hips so as to elevate the pelvis and allow the blood to run out of the piles. In some cases the application of an ice bag gives more relief than a poultice. This bag should be applied for ten or fifteen minutes at a time.

The best operative treatment, in Dr. Wright's opinion, is to take the little tumor between the thumb and finger, pass a curved, sharp-pointed bistoury down into the base of the tumor and cut your way out, following up the incision by a little compression Any hemorrhage which may follow is controlled by a bit of styptic cotton. Any cutaneous flabby tags of flesh remaining about the verge of the anus are taken up with a pair of dressing forceps and snipped off.

Internal Piles.—Every time they come down the patient is instructed to give them a free sponging with cold water. Where the bleeding is of an alarming character, and the patient does not care to submit to an operation, two ounces of a solution of one grain of the sulphate of iron to the ounce of water are injected into the bowels at night, and allowed to remain.

Another very excellent application which may be made with the finger is the following:

℞ Comp. ointment nut galls, ... ℥i;
   Aq. ex. opium, ... ℈i;
   Sol. subsulph. iron, ... ℥i;
Mix and make an ointment.

Dr. Wright strongly deprecates incision of intestinal piles. He either ligates or cauterizes them.

SACRO ILIAC DISEASE.

Dr. L. A. Sayre's treatment consists of leeches and warm fomentations, followed by ice-bags, extension and counter-extension, as in the treatment of fracture of the femur, upon the articulation, while the patient is in the recumbent position. When the acute symptoms subside, occasional blisters or the actual cautery are employed. The patient should now get the benefit of out-door exercise, which may be accomplished by increasing largely the thickness of the sole and heel of the foot of the unaffected side. This will lift the patient so that the diseased limb will swing clear of the ground, and this weight will generally be sufficient to give ease to the inflamed joint. If it is not, the weight may be increased by a leaden sole to the shoe of the affected side.

WOUNDS.

Dr. Brinton says that tannic acid, in powdered form, applied to wounds accompanying compound fracture, will convert them, when the wounds are not extensive or torn, into simple fractures, by rapidly forming a cicatrix, and thus save from one-third to one-half the usual time of healing.

CYSTITIS—A NEW TREATMENT.

Dr. Frank Hastings Hamilton suggests that horse-back riding is a most efficient method of treating this troublesome complaint. He has recommended this form of exercise in several cases in his private practice with the most advantageous results. This habit must be continued daily, for months at a time.

THE ACTUAL CAUTERY IN SCIATICA.

Dr. Ed. C. Janeway heats the platinum points to a white heat over an alcohol lamp, and then makes the slightest touch, very quickly, over the course of the nerve. His object is not to make a deep scar, but simply to produce a sort of blistering effect; a superficial inflammation, which, when it heals, leaves for a time a red line, but no permanent scar.

INTERNAL HEMORRHOIDS.

Dr. Hayes Agnew believes in the treatment by ligation. In thus ligating he always cuts a little grove in the mucous membrane round the base of the tumor, and allows the ligature to rest in this groove. In this way he saves his patients much pain, for it is this division of the mucous membrane that is the most painful part of the operation. When an anæsthetic is not employed the patient is made to sit over a vessel of warm water and to bring down the piles by straining.

When anaesthesia is produced, he draws them out with the forceps and seizes them with tenacula. He then passes his thread carefully through the base of the pile and withdraws the needle. He then takes hold of the two inside threads and having first cut a groove, ties them very tightly together. He is always careful to take out the tenaculum before tying the second set of threads, for otherwise the bleeding following the removal of the tenaculum would collapse the pile and so prevent the possibility of its being strangled. The piles are replaced in the bowel when tied. The bowels are closed for a week with opium and then a dose of oil is given.

FALSE ANEURISM OF THE THIGH.

Treatment in such cases may be either one of two alternatives. Dr. D. Hayes Agnew either amputates the leg at the thigh, or ligates the artery above and below the seat of injury, then exposes the parts fully and turns out the clots. In adopting the latter of these alternatives he rips open the whole length of the thigh over the site of the tumor, controls the bleeding rapidly with his fingers, turns out the clots, ligates the femoral artery above where it gives off the profunda and below the seat of the injury, and then inserts a good-sized drainage tube, bringing it out through an artificial opening made in the popliteal space. The parts are then brought together and a compress applied.

HYDROCELE.

"When we fail to effect permanent results by drawing off the fluid from time to time, if it continues to return, we will have to resort to the injection of some irritating fluid, as the tincture of iodine; or almost any other irritant of corresponding strength will answer the purpose. The object is to excite an inflammation which shall cause the two surfaces of the tunica vaginalis to be glued together, in short, to abolish the secreting surface, thus producing a radical cure. This injection is usually followed by a pretty sharp inflammatory action for a few days; about like that which we have associated with swelling arising from an epididymitis. After the inflammation subsides, if it has been sufficient, the hydrocele will not return. In some cases we make an incision down the entire length of the tunica vaginalis and pack it with oakum or carbolated lint to keep the wound open until it granulates from the bottom, and in that way destroy this secreting surface, and so produce a radical cure of the hydrocele."—*Dr. F. N. Otis.*

SCALP LACERATIONS.

When we speak of scalp wounds we divide them, clinically, into three classes; first, those in which only the skin and superficial fascia are involved; these are attended by very little displacement, simply from the fact that the dense connective tissue found beneath the skin binds it down to the aponeurosis of the occipito-frontalis muscle. These wounds are not of serious import, and readily unite. They can be brought together with adhesive plaster, or by merely separating a few hairs from opposite sides, and clamping them together with perforated shot. We sometimes use the twisted suture, especially if there be some bleeding, but usually the adhesive straps, or clamping the hairs are sufficient. Secondly, the wound may divide the occipito-frontal aponeurosis as well as the superficial structures, and this complication introduces a new feature into the case, because underneath the occipito-frontalis there is very loose connective tissue. You have noticed in dissecting that this aponeurosis can be removed with the handle of the scalpel or by the finger with little difficulty; it is very loosely connected with the parts below. When we have the wound extending to this structure, it is apt to be followed by much displacement. Moreover, when pus forms, it burrows in this loose tissue, there being no mechanical obstacle to its diffusion, and being imprisoned by the denser structures above, its tendency is to accumulate, and it may extend as far forward as the supra-orbital ridge. For this reason such cases require a great deal of watching. The first thing to be done is to secure any bleeding vessels, and some of these may be large enough to require a ligature. The three branches of the temporal artery found in this situation are often of sufficient size to require a thread. After removing with a stream of water, or soft sponge, any dirt or foreign substance between the lips of the wound and shaving the parts, the edges may be approximated, either by one suture of silver wire, or by the application of a number of adhesive straps. There has been great diversity of opinion among surgeons as to the propriety of introducing sutures into the scalp; it has been supposed that stitches encourage inflammation, and often lead to erysipelas: a result which, in my opinion, is not so much due to the stitches themselves as to the tension made in bringing the edges together, particularly where there is loss of structure. The wound should not be drawn tightly together by sutures, because when swelling occurs the tension is greater than the parts will bear, and may, therefore, lead to inflammation of an erysipelatous character. With this precaution I believe that silver sutures can be introduced into the scalp with as little risk as anywhere else in the body.

Examine the case frequently, and see that there is no accumulation of blood or pus beneath the flap. In addition to the adhesive straps, which should be applied in sufficient number to support the flaps in good position, there is another thing to be done in the present case in order to insure the accurate adaptation of the flap to the periosteum, namely, the application of a bandage and compress, so as to prevent the gravitation of the pus on the side of the head to the neighborhood of the ear. The compress is made of patent lint or muslin, and is secured by a few turns of the roller.

The third class of scalp wounds is one in which all the tissues are divided down to the bone. It is evident that these require even more care than the preceding, as there may follow purulent separation of the periosteum and exfoliation of bone. Of intra-cranial complications I will not speak. These scalp wounds unite with great rapidity in consequence of the decided vascularity of the part. A cold-water dressing is generally all that is needed.—*Dr. D. Hayes Agnew.*

HYDROCELE.

Dr. R. J. Levis began in 1872 to treat hydrocele with carbolic acid injections, because a more plastic grade of inflammation than that obtained by ordinary injections was required, and because incision gave rise to cure only thorough suppuration. His method is to withdraw the fluid by an ordinary trocar, and then to introduce the long nozzle of a syringe through the trocar into the vaginal sac. By this means the carbolic acid is thrown into the cavity, and there is no danger of its being injected into the cellular tissue of the scrotum. The carbolic acid crystals are merely liquefied by slight heat, or by a few drops of glycerin. To keep the injecting fluid ready for use at all states of temperature, about ten per cent. of glycerin or water may be added to the crystals. In summer the crystals become liquid without the addition of any solvent. The amount of carbolic acid which Dr. Levis injects is one-half a fluid drachm, and this is allowed to remain in the vaginal tunic. The operation is almost, if not entirely, painless, because of the local anæsthetic action of carbolic acid. The patients sometimes exclaim at the moment of introduction, but have a sensation of numbness rather than of pain. The pain, when tincture of iodine is employed, is much greater. Care should be observed to allow no acid to flow upon the external surface of the scrotum, for pain and inflammation will follow such contact. After the injection the patient is permitted to walk about the house until the weight and slight soreness of the scrotum cause him to lie upon a bed or lounge. The results after this method of treatment are excellent, for undue inflammation does not occur, there is no marked pain, and a radical cure generally occurs. Dr. Levis has never seen suppuration or sloughing follow this manner of dealing with hydrocele.

PRURITUS ANI.

Dr. Packard uses:

  ℞ Camphorae,
     Chloral hydrat., ... āā ℨss.
  M Unguent. petrolei, ... ℨvij.
Sig.—To be used as an ointment.

BURNS.

Dr. Shrady recommends that burns be treated by applying a paste composed of three ounces of gum arabic, one ounce of gum tragacanth, one pint of carbolized water (one part to sixty), and two ounces of molasses. The paste is to be applied with a brush, renewed at intervals, and is stated to be a successful method. Four applications are usually sufficient, the granulating surfaces being treated with simple cerate or the oxide of zinc ointment, as indicated.

COMPOUND DISLOCATION OF THE ANKLE JOINT.

In a late discussion, before the Philadelphia Academy of Surgery, Dr. Addinell Hewson stated that his experience was not altogether definite in pointing to the treatment by amputation. He had gained a good result, with only slight deformity, in a young person who had sustained a compound dislocation of the ankle He had reduced the luxation, and then suspended the limb while keeping it lying on the fibular side.

Dr. William Hunt's experience in the Pennsylvania Hospital among adults, led him, as a rule, to favor amputation. There was less risk in amputation than in conservatism, and a good point for the application of an artificial leg was attainable. If the shock would not permit a primary amputation, a secondary amputation might be available.

Dr. Thomas G. Morton had at times made an incision in order to get an opportunity to examine the parts, when he felt sure there was a fracture leading into the joint. In such cases he advised immediate amputation.

Dr. John H. Packard thought that each case should be treated upon its own merits and by no rigid law, for, though statistics showed that amputation at the lower third of the leg was favorable, the fact that the posterior tibial artery was intact would render conservative efforts justifiable in certain instances. The involvment of the tarsal bones in the injury was an important element to take into consideration, since the probable occurrence of subsequent necrosis would make the condition less favorable for conservative measures.

Dr. John Ashhurst, Jr, thought that several classes of injury were being confused in the discussion. When the astragalus was entirely dislocated there was much greater injury done to the soft parts than when only partial displacement occurred. Again, a fracture entering the joint made the case much more serious than a simple fracture of the fibula, though there might be a dislocation in either instance.

In Pott's fracture with partial displacement of the astragalus, accompanied by a wound, there is no true dislocation, and conservative treatment without operation may be tried.

If there had occurred a true compound dislocation without fracture, and operation was required, he would prefer in young and healthy adults excision to amputation. If, however, there was compound luxation with fracture of a bad form, amputation was to be done. The danger in all these compound injuries arises from confined pus, and free escape for pus is certainly given by excision. In his opinion, complete excision of the astragalus was better than to attempt partial excision of the joint. Statistics show that primary excisions are not so unfavorable as they were formerly considered. When the joint was completely crushed, amputation would be done by all surgeons.

Dr. J. Ewing Mears mentioned in this connection a patient who died from pyæmia, after having refused to allow amputation for an injury of the kind, where great comminution existed.

FRACTURE OF THE FIBULA.

As to treatment, in Jefferson Medical College Hospital the fracture-box is employed almost altogether; in those cases, however, where there is very great and uncontrollable lateral displacement, a Depuytren splint is employed, the foot being carried strongly inward. The great trouble with this plan of procedure is that the patient does not bear it well, as it is apt to be exquisitely painful. Consequently, in twenty-four or forty-eight hours, or as soon as this tendency to displacement is overcome, the limb is placed in the fracture-box and treated as a case of ordinary fibular fracture.

CLUB-FOOT.

In all cases the treatment, Dr. Morton believes, should commence in earliest infancy, with the view, at first, of correcting the varus, and consists in frequent daily manipulations of the foot and carefully-directed pressure on the tarsus by the hand of the mother or nurse. No apparatus is available at this tender age, and no tendon should ever be divided for the relief of club-foot in infancy, for a varus can, by careful and persistent stretching of the foot, always be overcome; only now and then the planter fasciæ must be divided. It is difficult and often impossible to reduce an equinus, but this deformity can also, by stretching, be much improved, and can be corrected by operation much better when the child is able to walk. Indeed, it is better to delay the section of the tendo Achillis until the child is two or three years of age, as the elevation of the heel during this time gives rise to but little inconvenience. If this tendon is cut early, a rigid contraction generally results. After the varus is corrected and the child is ready to walk, a brace is necessary to overcome the tendency to a recurrence of the deformity. The ordinary club-foot walking-shoe allows only of a hinge-motion, and as a frequent stretching of the ankle bone outward is desirable, Dr. Morton has long employed for this purpose a modification of the usual apparatus, as follows:

Taking an ordinary leather shoe, which should lace up in front, with the lateral steel supports running up above the middle of the thigh, with transverse braces and bands above and below the knee to hold the apparatus in position, he has had an additional hinge placed opposite the external malleolus, and opposite this point a portion of the inner steel rod has been taken out, and replaced by a double antero-posterior hinge or toggle-joint, which enables it to yield when pressure is made upon it, while the hinge in the outer support allows the turning out of the foot each time it is brought down to the ground. The weight of the body resting upon it with each step gives, by this means, an outward movement or partial rotation of the foot, which is consequently communicated to the tarsus, so that this portion of the articulation is made more pliable.

In children where operations have, either through neglect or ignorance, resulted in stiff and unyielding deformity, and in adults never subjected to treatment, it becomes necessary after the division of tendons and fascia to stretch the foot at once into position by powerful pressure. For this purpose Dr. Morton has invented an apparatus which he has used in conjunction with manipulation, consisting of bands and screws, which can be applied first under anæsthesia, in order to force the condensed and rigid tarsus and surrounding tissues into a normal position. Other forms of apparatus for a similar purpose he has employed with the same object in view, and with their aid has been able to accomplish very much. He has never seen a case where excision of the tarsal bones was required. It is quite probable that in some rigid adult cases the tarsal bones have been fractured, however, by the power employed with the stretcher.

After stretching or tenotomy he always uses for some days a posterior well padded tin splint, and has the foot dressed morning and evening, carefully avoiding any irritation of the skin, and says that the only forms of talipes which give trouble are equinovarus and varus, hence he confines his remarks to these varieties of club-foot.

TO REMOVE FOREIGN BODIES FROM THE CORNEA.

"Take a little stick, place upon one end of it a small pledget of cotton, and, after having moistened this, brush it over the surface of the cornea at the point where the foreign body is, and it will very likely be caught in the fine meshes of the cotton. Having prepared such a probe, the eyelid is drawn up against the superciliary ridge and the eyeball kept from rolling. Then, seeing exactly where the foreign body lies, you quickly brush over the surface, so impinging the cotton against the surface as to cause the foreign body to have a tendency to rise into it. In this way your may remove all foreign bodies not deeply imbedded. If you do as is so often done, namely, pick at the foreign body with a cataract needle, you may not get it out, and by picking you may cause a great deal of pain. So I would recommend you to follow the directions which I have just given, which will answer in all ordinary cases. Suppose, however, that the foreign body is so deeply imbedded that you cannot get it out by means already cited, you can then take a cataract needle or a Bowman's spud so flattened at the extremity that it may be passed under the foreign body and tend to lift it out. You must be exceedingly careful not to injure the cornea, for that will impair the vision Suppose there is a little mass of steel imbedded in the cornea, and you can see that it has gone through the cornea, so that the point of the steel is in the anterior chamber, and when you try to remove it you see a little leakage of the aqueous humor, then what should you do? You must get something back of the foreign body, or it will fall into the anterior chamber. You can do this by means of a Beers knife and under the influence of an anæsthetic. The eyeball should be steadied with a pair of fixation forceps, and the knife placed about a line from the body and perpendicular to the cornea. It should be carried right through the cornea into the anterior chamber, and then the blade of the knife turned and the handle depressed, so as to bring the point of the knife out beyond the foreign body, thus preventing it from falling into the anterior chamber. Thus you have the foreign body lying held in the cornea. It lies spitted upon the knife. I think I have been enabled by this method of procedure to pick out more than one foreign body and save the patient's eye.

Now, I do not think that eye-stones are of any value in removing foreign bodies from the eye. The only way they act is this. They are disk-like masses which tend to pass over to the inner canthus of the eye and there to discharge themselves. The advantage that the eye-stone is supposed to have is that it elevates the lid and allows the tears to wash the foreign body off from the surface of the eye. That might be an advantage in cases where simply a cinder got behind the upper eyelid, but this is of no value in cases where the foreign body is imbedded in the cornea. It is a very bungling method, and I never use it. Now this man here will get well in a few days, I think. I would recommend that he keep the eye quiet and protect it as well as possible. The only thing I would use in the way of a wash would be a little table salt in a little warm water, say a teaspoonful to the pint.—*Dr. Cornelius Agnew.*

TREATMENT OF VESICAL CATARRH BY ESTABLISHING URINARY FISTULÆ.

In hopless cases of chronic cystitis it has occurred to Dr. D. Hayes Agnew that the life of the patient might be made comfortable by separating the connection of the ureters with the bladder and bringing them out through the abdominal walls, establishing fistulæ either in the iliac or in the lumbar region, and thereby diverting the urine entirely from the bladder. That such a route for the escape of the urine is not so objectionable as might be supposed will appear from the experience of two persons in Philadelphia who suffer from urinary fistula occasioned by accident, one of whom is able to attend to his occupation—that of a daily laborer—by swathing his body with a thick roll of bandage, by which the urine is absorbed. If the fistulæ were favorably situated, mechanical appliances might be constructed in which to receive the urine.

The feasibility of the procedure proposed has been satisfactorily verified by dissection and operation on the cadaver. At first it was supposed that the proper route to the uterus would be through the loin, as in lumbar colotomy; but the colon on each side is an obstacle which cannot readily be overcome. The plan which Dr. Agnew pursued was to make an incision beginning one inch below the anterior extremity of the last rib, and terminating two inches below the anterior superior spinous process of the ilium. After dividing the skin, superficial fascia, external and internal oblique and transversalis muscles, the transversalis fascia is next broken up, together with the loose tissue connecting the peritoneum with the iliac fossa. It only remains to detach carefully the serous sac until the primitive iliac vessel is reached, at the bifurcation of which into external iliac and internal iliac the ureter will be found to pass into the pelvis.

Following the tube down, it should be severed as near to the bladder as possible, two ligatures having been previously applied (the lower one catgut), and the division made between the two threads. To relieve any tension on the ureter, a puncture is next made through the parietes a short distance above the upper angle of the wound, and the urinary duct piloted through by means of a probe secured to the end of the ligature previously attached to the ureter. It only remains to detach the thread from the duct and to secure the latter by two stitches to the external opening, after which the main wound can be closed. It would not be proper to operate on both ureters at the same time. The patient should be allowed to recover from the first before proceeding to the second. Nor would such a surgical procedure be advisable if there was reason to believe that the kidneys were seriously implicated.

POLYPUS OF THE RECTUM.

Follicular Polypus.—Dr. Williston Wright's treatment is exceedingly simple Thus, when the pedicle is very small and contains no large vessels, it is only necessary to seize it with a small pair of Vulsellum forceps or any other instrument that will hold it securely, and twist it around in one direction until the pedicle breaks off and the tumor comes away; treating it, in other words, as we commonly treat a polypus of the nose. On the other hand, if the polypus is vascular, and especially if you can feel a large artery passing through its centre, it will be safer to lift the tumor away from its attachment with the Vulcellum forceps and apply a soft flat ligature close down to the origin of the pedicle. The ligature should be tied only tight enough to arrest the circulation. If you go beyond this point, or use a round, hard-twisted piece of silk, there is danger of cutting the pedicle off at once, and of having the same amount of hemorrhage as would have resulted from a use of the knife or scissors. The best ligature for this purpose consists of linen tape about one sixteenth of an inch in width.

The bowels should then be confined for about two days by giving one or two small doses of opium, and at the end of that time a mild aperient will probably bring away the polypus and the ligatures together.

Fibrous Polypus.—Dr. Wright's treatment consists in tying a round, firm ligature tightly around the pedicle, close down to the mucous membrane. Now mark the difference in treatment in the two forms of polypus. In the first variety he advised a soft, flat ligature applied rather loosely. In this form you are to use a round, well-twisted silk cord, and apply it as tightly as possible. The fibrous polypus is so much harder than the follicular variety that unless tied thoroughly the circulation will not be controlled and the tumor will be a long time in sloughing off, to say nothing of the danger of secondary hemorrhage.

If the base of the tumor is too large to be divided in this way, then it had better be encircled with the chain of the ecraseur and gradually crushed off. To do this satisfactorily, however, the patient must be etherized and the tumor thoroughly exposed by the use of a proper speculum, since in a few cases, whatever care is used in crushing off the pedicle of the tumor with this instrument, there will be more or less hemorrhage, and you will be obliged to use the actual cautery or some powerful styptic, or possibly to take up a large vessel and submit it to torsion or secure it by the use of the ligature.

FRACTURE OF THE VERTEBRÆ.

The patient should be placed in the best possible condition to avoid bedsores, which, owing to defective innervation of the integument, are very apt to supervene, becoming at times so large that the patient is unable to bear up under the tremendous sloughing, and so finally sinks a victim to them. The best way to avoid this disaster, according to Dr. R. J. Levis, is to place the patient on a water-bed, which makes equable pressure on all parts of the body, and not on the salient parts only (like the buttocks, shoulders, etc.), as the ordinary mattress does. While the rubber bag makes the best and most convenient water-bed, yet the expense attached to its purchase is frequently too great for the patient's means. He has devised a plan by which this objection is to a large extent obviated, while at the same time all the good effects of the bed are obtained. By making a water-proof trough, just large enough to allow the patient to lie in it comfortably (allowing some margin for turning him over, etc.), filling it with water and stretching loosely over it an ordinary gum blanket (costing from $2 to $4), this being made just tight enough to comfortably float the body, a very nice substitute for the more expensive article is obtained. The calves of the legs and the heels must be carefully watched and guarded; these and the other parts of the body liable to bed-sores being daily bathed and thoroughly dried. The bowels should be carefully attended to, as dribbling of fæces is apt to take place, producing great irritation of the integument. The bladder should be catheterized and kept thoroughly washed out, for the urine in these cases is extremely liable to become decidedly alkaline and phosphatic, and these deposits accumulating in the bladder the urine becomes more or less fetid. To deodorize the water and wash the walls of the bladder, he is in the habit of using a saturated solution of boracic acid, which, being very mild, can produce no evil consequences; at times, also, he employs an injection of carbolic acid. In case of dribbling a urinal, fitting between the patient's limbs and maintained in position, should be employed.

MEDICAL AND SURGICAL DISEASES OF WOMEN.

CURVED SPONGE-TENTS FOR UTERINE FLEXIONS.

According to Prof. Ellerslie Wallace the first tent which is to be used in attempting to erect a flexed uterus should be of small size; it should not contain a spring, because the elasticity of the spring will straighten to some degree the small tent, the bulk of the sponge not being sufficient to hold the spring down in its proper curve.

As a general rule, tents of three curves will be all that will be found necessary for ordinary cases, and therefore a number may be manufactured at one time for future use. The shapes most used are, first, a moderate curve, only a little more than the natural bend of the hand; secondly, a fish-hook curve, for extreme cases; and, thirdly, an intermediate one, which will probably be the most often required. After insertion, the spring is fastened by a silk ligature passed through both it and the sponge, the needle in its passage traversing obliquely the aperture in the spring, then carried around the cylinder—one quarter of its circumference—and passed again through the sponge and spring; finally, the ligature is drawn tight, and tied at the point of entrance, burying it deep in the sponge. Instead of carrying the thread around on the outside of the cylinder, it may be passed under its surface by taking a stitch through the sponge, making a "subcutaneous" ligature. Transfixing the other end of the tent and the second aperture in the spring by a stout needle, the spring is perfectly secure.

A very thick solution of gum arabic is required, in which the still moist tent must be thoroughly soaked. It is now to be taken out, firmly wrapped, like thread on a spool, with strong twine, from one end to the other, and back again. The tent is now ready to be moulded into any desired curve. Having a sound bent to the shape of the uterine cavity, is laid down upon a piece of wood, and its course indicated by several tacks. The sound is now replaced by the moist tent, which is allowed to remain until it dries, the position of the spring being indicated by the needle at its inferior extremity. The hard and dry tent is next taken out, and the twine removed. Its surface presenting a rough appearance from the indications produced by the cord, it should be lightly smoothed with fine sand-paper, and the point somewhat bevelled; but the tent should *not* be made to gradually taper to a point—as most tents are—because it is next to impossible to retain such a wedge-shaped instrument in the uterus. The tent may now be rubbed with a little wax, and burnished with any hard substance; the handle of a pair of scissors answers very well.

Finally, a string may be passed through the opening left by the needle in the lower end, for convenience of extraction and to secure the end of the spring in the centre of the sponge.

INFLAMMATION OF THE NIPPLES.

This formulæ is highly commended by Dr. Albert H. Smith as a protective application:

R Emplast. plumbi, ... ℨij;
Aetheris sulphurici, ... f ʒss;
Flexile collodion, ... f ℨj.

Sig.—Powder the lead plaster, add the ether, and mix them well together before adding the collodion. It makes a creamy mixture, and is to be applied with a brush over every portion of the carefully dried nipples, with the exception of the openings of the milk ducts.

PRURITUS VULVÆ.

In the treatment of this disease Dr. Goodell always strives to discover the causes. Often these are obscure, and, even when found, it is necessary to treat the itching itself, apart from the cause.

So intractable is this disease that it is well to have a list of approved formulæ for the exacting symptom of itching. Three domestic remedies of value may first be tried. With water as hot as can be borne, the vagina should be syringed out, and the external genitals bathed several times a day. Striking the external genitals with a sponge wrung out of boiling water, will frequently give the greatest relief. An infusion of tobacco, one ounce to the quart of boiling water, may next be applied in the same manner. This failing, the itching parts can often with advantage be washed with cider vinegar, either diluted or of full strength. Decoctions of walnut leaves, of flaxseed, of quince seed, of slippery elm bark, or of sassafras pith, are very soothing remedies, if not positively curative. These applications may be medicated with the zinc sulphate, lead acetate, or with borax.

The following formulæ have been prescribed by Dr. Goodell with great benefit. Some of them are his own, others are the gleanings of years, and are, so far as he knows, without ownership:

℞ Aluminii nitratis, ... gr. vj ;
   Aquæ destillatæ, ... f ℥ j. M.
Sig.—Apply with a soft sponge.

℞ Iodoformi, ... ℨ j ;
   Balsami peruviani, ... f ℥ j. M.
Sig.—Smear the parts with a brush.

℞ Chlorali,
   Camphoræ, ... āā f ℨ iv.
Rub these into an oil ; then add
   Unguenti simplicis, ... ℥ j ;
   Pulv. acidi boracici, ... ℨ iv. M.
Sig.—Apply with a brush.

℞ Acidi acetici, ... f ℥ j ;
   Glycerinæ, ... f ℥ iij. M.
Sig.—Apply locally.

℞ Acidi carbolici, ... gr. xij ;
   Morphiæ acetatis, ... gr. viij ;
   Acidi hydroc. dil., ... f ℨ ij ;
   Glycerinæ, ... f ℥ j ;
   Aquæ, ... ad. f ℥ iv.
Sig.—Apply locally.

℞ Ung. hydrarg. nitrat.,
    Olei morrhuæ, ... aa ℨ j. M.
Sig.—Anoint the parts twice daily.

℞ Chloroformi, ... f ℨ j;
    Olei amygd. exp., ... f ℨ vij. M.
Sig.—Apply to the itching parts.

℞ Sodii boratis, ... ℨij;
    Morphiæ muriatis, ... gr. xx;
    Acidi hydroc. dil., ... f ℨj;
    Glycerinæ, ... f ℨj;
    Aquæ rcsæ, ... ad. f ℥viij. M.
Sig.—Apply with a soft sponge.

℞ Potassii cyanidi, ... gr. j-iij;
    Liquoris calcis, ... f ℨiv;
    Adipis, ... ℨiv. M.
Sig.—Apply locally.

℞ Sodii bisulphitis, ... ℨvj;
    Aquæ . . . . . . f ℥vj. M.
Sig.—Apply with a soft sponge.

℞ Hydrarg. chlor. corrosivi, ... gr. j;
    Pulveris aluminis, ... gr. xx;
    Amyli, ... ℨiss;
    Aquæ, ... f ℨvj. M.
Sig.—Apply locally.

In desperate cases electricity may be applied directly to the itching vulva and vagina; or the parts may be cauterized with the solid stick of the silver nitrate. Very recently good effects have been reported from repeated hypodermic injections into the moist itching points, of from fifteen to twenty-five minims of a two per cent. solution of carbolic acid. In every case it is well to keep the labia apart by a pledget of lint soaked in one of the preceding lotions. When it is clear that the pruritus depends upon a uterine secretion, such as a senile leucorrhœa, the discharges should be neutralized before they reach the vulva. This is best done by dipping a tampon of cotton-wool into a medicated glycerole, such as that of tannin, of morphia, of lead, or of zinc, and pushing it up against the cervix. Scanzoni uses for this purpose a tampon of cotton-wool thoroughly sprinkled with equal parts of finely powdered sugar of alum. To secure sleep the potassic bromide, chloral, or opiates, in full doses, must be resorted to. When due to diabetes, the itching will be greatly allayed by twenty-grain doses of sodium salicylate given by the mouth thrice daily. For this valuable remedy Dr. Goodell is indebted to Dr. James Simpson, of Philadelphia, who first brought it to his attention.

When pediculi or other insects are the cause of the pruritis, there is no surer remedy than a liniment composed of one part of carbolic acid and nine of sweet-oil.

"This exceedingly obstinate disease I have often permanently relieved if caused by an erythema, intertrigo, acne, eczema, or prurigo produced by vaginitis or endometritis. In the former, by warm injections of flaxseed tea with a solution of the aqueous extract of opium, together with a sitz-bath, lukewarm, twice daily for twenty or thirty minutes, and the subsequent irrigation of the vagina by means of a fountain syringe and adding from one-half to a teaspoonful of sulphocarbolate of zinc to the quart of water, used every two hours while the patient lies in the horizontal position with her hips well raised. After each vaginal irrigation a tampon of carbolized or salicylated cotton, with some unguent. plumbi and belladonnæ, is introduced into the vagina so that it prevents the external parts from being bathed in the secretions, often mucous, sometimes muco-purulent—which cause the excoriations at and around the vulva, the nates, and the inside of the thighs—and directing the patient to use in the evening, and also for one hour in the afternoon, applications of black wash—aquæ phagedenicæ nigræ. Where the pruritus or prurigo disturb the patient's sleep, a hypodermic injection of morphia is given. The pruritus itself is greatly relieved, besides the above medication, by application of dilute tincture of iron or a twenty per cent. solution of carbolic acid. The most useful of all applications, however, for pruritus vulvæ, I have found the application of the balsam of Peru, of which I use the following:

℞ Pulv. gummi arabic, ... ʒij.
    Peruvian balsam, ... ʒj.
    Oil of Almonds, ... ʒjss.
    Rosewater, ... ʒj. M.

Sig.—Apply freely with a camel's-hair brush, eight or ten times a day, to the itching part.

"This latter prescription, which has been first suggested by Hufeland, I have also used for the past nineteen years, with the most happy results, for sore nipples, applied every hour for a few days. It has never failed in my hands to cure this troublesome affection. I prefer this prescription to the use of borax and alcohol, the nitrate of silver, the zinc ointment, or the saturated solution of borax, highly recommended by others.

"If the pruritus vulvæ is dependent upon diabetic urine, I have found, in addition to the means herewith recommended, the daily internal use of from six to eight drachms of glycerine, in teaspoonful doses, extremely beneficial. If dependent upon granular vaginitis, I am in the habit of touching each granule, after first scraping it off with the curette, with an exceedingly fine point of nitrate of silver.

"The endometritis, if this is the cause of the pruritus vulvæ, is treated with laminaria or the curette; if dependent upon polypoid or fungoid or adenomatous multiple growths within the cervical canal or the cavity of the uterus, after which, if the whole uterine cavity has been dilated thoroughly so as to admit the finger to the fundus uteri, which generally takes, by means of laminaria, about thirty-six hours, with Churchill's tincture of iodine applied every three to four days; or, if there is much sensitiveness, I apply carbolic acid drachma semis to the ounce of glycerine, upon the flexible uterine applicator, down to the fundus, leaving it in contact from five to ten minutes; or I irrigate the uterine cavity through a No. 6 male flexible catheter with a three per cent. warm solution of carbolic acid, always taking care, of course, not to make any intra uterine application unless the canal be thoroughly dilated, and that there is no inflammation in the substance of the uterus or its adnexa.

"I have, within the last nineteen years, thoroughly tested the medication herewith recommended, and I have yet to see a case where I have failed to permanently cure the most obstinate cases of either pruritus vulvæ, vaginitis, or endometritis, and the sore nipples which often undermine the health of mostly nursing women with very delicate skins.

"In communicating these notes I claim no originality in the application of the medication recommended. The treatment has proved highly satisfactory in my gynecological practice, and I desire to recommend it to thousands of practitioners who have not tried them.

RUDOLF TAUSZY, M. D.

PUERPERAL SEPTICÆMIA.

The treatment pursued at the Philadelphia Hospital consists in the early administration of gr. ss. of morph. sulph., the *immediate* application of twenty foreign leeches over the hypogastrium, sixteen to twenty grains of quinia daily, warm fomentations, abundance of liquid nutriment, stimulants and of intra-vaginal and intra-uterine, syringing with a solution of the permangante of potassium, of carbolic acid, or of Labarraque's solution.

PUERPERAL MALARIAL FEVER.

Dr. Fordyce Barker has found half ounce doses of Warburg's tincture given every four hours until the fever is abated, and then gradually lessened, more effectual than quinia in the treatment of this complaint.

INFLAMMATORY DISEASES OF THE GLANDS OF THE FEMALE URETHRA.

The treatment which Dr. A. J. C. Skene employed at first was to inject the tubules with the ordinary solutions used in the treatment of inflammation of mucous membranes, using for the purpose a hypodermic syringe, with the point of the middle rounded off. This method he found useful, but very tedious. It then occurred to him that laying open the tubules their whole length and keeping them open would prevent the purulent accumulation (which acts so effectually in keeping up the inflammation) and also bring the affected parts within easy reach of the necessary treatment. He has tried the method in quite a number of cases and found it entirely satisfactory. In the majority of cases it is all that is required to effect a complete cure. The method of operating is as follows: The patient is placed upon the left side, and a Sims speculum used to keep the labia apart and retract the perineum. This brings the parts well in view, and within easy reach of the operator.

The position and depth of the tubules having been first ascertained, the probe-pointed blade of a very fine scissors is then introduced, and the posterior wall divided its whole length. To prevent the parts from reuniting, a small piece of cotton, saturated with persulphate of iron, should be packed in between the divided edges. Brushing the surfaces over with the iron, without using the cotton, will answer, although less certainly to prevent re-uniting. Very little after-treatment is required. In the majority of cases recovery follows the operation of laying open the canals. Sometimes the inflammation lingers in a modified form, but yields to a few applications of nitrate of silver or sulphate of zinc. In several cases in which the excrescences were abundant, they remained after the operation, although very much reduced in size. An application of nitric acid destroyed them, and they have not shown the least disposition to return.

THE TREATMENT OF UTERINE FIBROIDS.

The indications to be met in the treatment of fibroid tumors are four in number: (1) to stop the bleeding; (2) to mitigate the pain; (3) to stop the further development of the tumor, and (4) to remove it if possible, entire, or, in case its size forbids extirpation, to get as much of it away as possible. The stoppage of hemorrhage is often the most important item at first. With this object in view Dr. Goodell puts his patients to bed, keeps them absolutely quiet, and administers ergot in large doses. It is his habit to give at least a teaspoonful of the fluid extract every three or four hours during the continuance of the bleeding. The combination of the oil of erigeron with the ergot is very desirable in some cases; if this treatment fails to do good, he uses vaginal injections of hot water and hot applications to the spine; or, next to ergot, he tries gallic acid in doses of twenty grains every second or third hour, until several doses have been taken. Opium either by suppository or internally is given in a sufficient quantity to relieve the pains. Hypodermic injections of morphia are sometimes needed. When he has once succeeded in stopping the hemorrhage, he takes some precautions during the shortened intermenstrual period to prevent its recurrence. This is accomplished by giving medicines that reduce the congested state of the womb and allay its nervous excitement. Bromide of potassium, tincture of nux vomica and iron are prescribed for this purpose. Iron is to be given cautiously and only when the patient happens to be anæmic, for iron causes a tendency of blood to the womb, and so keeps up in a measure, the bleeding. In some instances quinia seems to have the same effect. To stop profuse and exhausting hemorrhage he is accustomed to set to work as follows: He takes a strip of lint and wets one end with Monsel's Solution. He then carries this end up to the crevix through a speculum, and packs it into the cervical canal by means of a uterine sound. The other end of the strip he allows to remain free in the vagina so that it can be removed easily. This packing of the cervical canal corks up the womb, and is sure to stop the drain of blood. Sometimes, instead of this lint, he introduces a clean sponge, or, what is better still, a sponge tent. The tent is preferable on account of its cleanliness. The blood oozes through it and washes it sweet and free from fetid discharges. The tent may be allowed to remain *in situ* much longer than the lint. His experience has proven, too, that a large sponge tent dilates the mouth of the womb, and, in some unexplained way, diminishes the bleeding. Various explanations of this fact have been proposed. Some hold that by pressure it diminishes the amount of blood in the cervical vessels, from which the bleeding may possibly come. The same result (i. e. stoppage of the bleeding) may often be accomplished by simple incision of the cervix. This incision should also be performed a few days before the expected appearance of the menstrual flow. Such dilations and incisions are sometimes followed by permanently good results.

When dilatation fails there is a resource still left in the division of the capsule, which will cut off a large source of blood, and will aid the efforts of the womb in extruding the tumor. He is in the habit of performing this operation with Kuchenmeister's scissors. The hook on one blade of this instrument prevents them from slipping or sliding backward while cutting. While it is best to make as large an opening into the capsule as possible, a small one will often suffice. He has known of a large tumor working its way out of an incision barely admitting his index finger. As the tumor emerges from its capsule, its extrusion is aided by traction and by severing its adhesions with the finger as they come within reach. When the fibroid is of the sub-mucous variety, he is in the habit of removing it by avulsion. This plan, as laid down in the books, is to seize hold of the prominent part of the tumor with the Vulsellum forceps, and, by both traction and twisting, to tear it off from its mucous capsule and out of its bed. A better way, he thinks, is to seize the growth, as before, with the Vulsellum, and pass up the wire-loop of an ecraseur over the tumor, and cut through its mucous capsule flush with the uterine wall. Traction is then made by both Vulsellum and ecraseur, and the growth is enucleated, without any assistance, from its envelope. The large opening will be closed up by the subsequent contraction of the womb. In this way quite a number of partly intestinal and partly submucous fibrous tumors have been removed with but one death, and that was from the occurrence of heartclot on the sixteenth day after the operation. Whenever surgical procedures are out of the question in these cases we can do no better than imitate the natural course of the tumor and help it to become a polypus. This end he accomplishes by means of full doses of ergot. The firm contraction of the uterine muscles will also lessen the supply of blood, and the tumor may diminish in size, and sometimes disappear entirely. The hypodermic administration of ergot is that which he follows. He has seen a number of cases greatly benefited by it. In one case the tumor almost wholly disappeared. He prefers Bonjean s solution of ergotin for the hypodermic needle. Next to this the best is perhaps the aqueous infusion, each minim representing one grain of the drug. The proper place to make the injection is in front of the abdomen just below the navel.

THE TREATMENT OF HYSTERICS.

When Dr. Goodell is called in to treat a young girl with an hysterical attack he (1) Institutes at once a firm pressure in the neighborhood of both ovaries. This is very apt to quiet the patient at once. (2) Admisters an emetic. He has found that a woman who is well under the action of an emetic has not the opportunity to do anything else than be thoroughly nauseated. [He gives a full dose of ipecac, with one grain of tartar emetic.] (3) [And this method of controlling the spasm will often act charmingly] he takes a good-sized lump of ice, and presses it right down upon the nape of the neck. This produces quiet by its powerful impression upon the whole nervous system.

When the attack is entirely under control the best method of preventing the occurrence of another attack is to administer a full dose of assafœtida—none of your small, two or three grain doses, but ten grains, all at once.

COCCYGODYNIA.

All anal and uterine lesions are remedied. Should no good follow, Dr. Goodell tries local injections of morphia or of carbolic acid, or rectal suppositories of iodoform. Where the case still resists treatment, surgical interference becomes necessary. This can be afforded in two ways. By one, the coccyx is cut down upon and extirpated by the bone forceps. By the other a tenotomy knife is passed in near the tip of the coccyx, and carried up to the articulation. It is then made to shave off from the bone all its muscular and tendinous attachments. The exposed tip is then lifted up by the finger, while the attachments are snipped off on every side by means of a pair of curved scissors.

HERNIA OF THE LABUIM MAJUS.

The treatment pursued by Dr. Goodell is like that for an inguinal hernia. The hernia is reduced by taxis and position, and the inguinal canal and the canal of Ruck are closed by the pressure of an appropriate truss. In obstinate cases, Dr. Goodell would feel tempted to perform an operation looking towards the permanent obliteration of this canal.

ENCYSTED HYDROCELE OF THE ROUND LIGAMENT.

Dr. Goodell generally effects a cure by emptying the hydrocele either by means of the hypodermic syringe or with the finest needle of the aspirator, and by injecting into the sac and leaving there a few drops either of the tincture of iodine, or of liquid carbolic acid.

PUDENDAL HEMORRHAGE.

Such bleeding is met by the application of lumps of ice, and by injections of a saturated solution of alum into the wound. The solution of alum is perhaps not so astringent as that of the sulphate of iron, but Dr. Goodell claims that it does not give rise to those plaster-like clots, which become putrid before they can break down, and thinks it altogether a cleaner remedy. Should these fail, the vagina is packed firmly with oakum, or with cotton wool, and a compress is laid on the wound and kept there by a T bandage. Should the hemorrhage continue the whole thickness of the labium is nipped with a pressure forceps, or with a spring clothes-pin. If the worst comes to the worst the bleeding vessels, however deeply situated, are compressed by silver sutures passed beneath the wound and from the cutaneous to the mucous surface.

PUDENDAL HAEMATOCELE.

These blood tumors are best treated by rest and by cold applications. If not large they may undergo absorption and wholly disappear. Should the coagulum liquify or break down, Dr. Goodell makes a free incision on the mucous surface and turns out the clots. He afterwards washes out the cavity daily with carbolated water until all danger from septicæmia has passed away, and the healing process has become fully established. If a secondary hemorrhage occurs it is treated in the same way as a primary hemorrhage.

HEMORRHAGE FROM THE HYMEN.

Dr. Goodell arrests hemorrhage of this nature by stuffing a large sponge into the vulvar opening.

INDEX.

A.
Abscesses and Ulcerating Surfaces..................69
Acne..........................29
Anæmia, The, of Children...24
Anasarca......................18
Aneurism, Electrolysis in Treatment of..............22
Ani, Pruritus.................84
Ascaris Vermicularis and Round Worm..............59
Asthma........................42

B.
Bronchitis, Acute, in Children..55
Bubo..........................72
Burns.........................84

C.
Catarrh, Nasal............7, 33
" Vesical................90
Cholera Morbus...............16
Club-Foot....................86
Coccygodynia................105
Constipation.............29, 63
Convulsions in Children......26
Cornea, Foreign Body in the..88
Coryza, Fœtid................17
Cough, Nervous...............61
Cystitis..............12, 76, 79
Cystorrhagia.................67

D.
Dermatitis...................18
Diabetes Mellitus.............7
Diarrhœa, Serous, of Infants..21
Diphtheria...................42
Dropsy, Cardiac..............49
Dysentery, Croupous..........53

E.
Eczema.......................28
Effusion, Pelvic.............15
Endarteritis.................41
Endocarditis, Rheumatic......19

F.
Fever, Delirium of Typhoid..20
" Diarrhœa of " ...15
" Intermittent...........10
" Intestinal Hemorrhage in Typhoid..........20
" Malarial...............17
" Parotid Swelling of Typhoid............32
" Pneumonia of Typhoid.19
" Puerperal Malarial....100
" Scarlet.................26
" Sore Throat of Typhoid.13
" Typhoid................13
Fracture, Potts'.............70
" of Fibula..........86
" " Vertebræ.......92

G.
Goitre, Exophthalmic.........33
Gout.........................28

H.
Hæmatocele, Pudendal........106
Heart, Fatty.............27, 64
" Flagging...............16
Hemorrhage...................55
" Pulmonary.........5
" Pudendal.......106
Hemorrhoids..............76, 79
Hydrocele................80, 83
Hymen, Hemorrhage from the.........................107

INDEX.

Hypertrophy................63
Hysteria...................104

I.
Incontinence of Urine.......37
Indigestion..................9
Iritis......................62

J.
Jaundice, Obstructive........30
Joint, Caries of Ankle........75
" Compound Dislocation of Ankle.................84

K.
Kidneys, Floating............22

L.
Labium Majus, Hernia of the.105
Lids, Granular..............32
Ligament, Hydrocele of the Round..................106
Liver, Sclerosis of the........42
Lumbago....................36

M.
Malarial Attacks, Obstinate...34

N.
Nipples, Inflammation of the.95

P.
Pediculosis Capitis...........29
Phthisis, cough of........11, 52
" Diarrhœa of........54
" Fever of............62
" Gastic Disturbances of Initial...........59
" Hæmoptysis of......57
" Laryngitis of........56
" Night Sweats of.....61
Pleurisy, The Cough of.......57
" and Empyema in Children............33
Pneumonia..................20

R.
Rectum, Polypus of..........91
Rheumatism, Acute....5, 31, 53

S.
Sacro-Iliac Disease...........78
Scalp Lacerations............81
Sciatica..................19, 79
Sclerosis, Multiple...........41
Scrofula....................71
Seborrhoea Corposis.........18
Septicæmia, Puerperal......100
Spina-Bifida.................67
Sprains, Old................73
Stomach, Catarrh of the.....37
" Dilitation of the....59
Syphilis....................71
" Hygiene of the Mouth in................66
" Infantile............68

T.
Tape Worm..................23
Thigh, False Aneurism of the.......................80
Thrombosis, Portal...........5
Tinea Sycosis...............17
Tonsillitis, Acute.............9
Typhlitis...................51

U.
Ulcer, The Chancroidal......74
Urethra, Disease of Glands of Female................100
Uterine Fibrids.............101
" Flexions............94

V.
Vertigo, Gastric.............22
Vulvae, Pruritus.............95

W.
Wounds, Slight..........69, 78

www.ingramcontent.com/pod-product-compliance
Lightning Source LLC
Chambersburg PA
CBHW020233240426
43672CB00006B/509